THE SOUTHWEST

Native Plant Primer

THE SOUTHWEST
Native Plant Primer

235 Plants for an Earth-Friendly Garden

JACK DASH, LUKE TAKATA
& THE ARIZONA NATIVE PLANT SOCIETY

Timber Press
Portland, Oregon

Frontispiece: Cacti, agaves, and stone act together to create an inviting entrance to a garden.

Opposite: A child harvests the fruit of Texas mulberry (*Morus microphylla*). Introducing children to the wonders of the natural world early on plants a seed that will germinate in the next generation of Earth-centered gardeners.

Page 6 (clockwise from top left): Acer grandidentatum, Asclepias linaria, Senna covesii, Maurandya antirrhiniflora, Aristida purpurea, Cylindropuntia imbricata

Timber Press
Workman Publishing
Hachette Book Group, Inc.
1290 Avenue of the Americas
New York, New York 10104
timberpress.com

Timber Press is an imprint of Workman Publishing, a division of Hachette Book Group, Inc. The Timber Press name and logo are registered trademarks of Hachette Book Group, Inc.

Printed in China on responsibly sourced paper
Text design by Mary Velgos based on a series design by Debbie Berne
Cover design by Sara Isasi based on a series design by Amy Sly

ISBN 9-781-64326-333-5
A catalog record for this book is available from the Library of Congress.

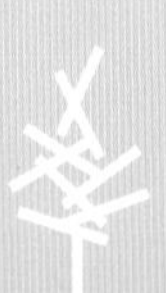

To Lano, Julian, and Reign
for giving us a reason to grow a better world, and to all the members of The Arizona Native Plant Society.

Contents

Preface

> Why should the Southwest be ashamed of the things that are its own? Why should we not take pride in the things with which nature has so richly endowed us?
>
> —*Forrest Shreve, American ecologist*

A gardening revolution is underway. The dominance of the manicured lawn and the barren patch of rocks is giving way to a desire for diverse, functional, beautiful urban landscapes that connect our yards and public spaces to the wider ecosystems around us.

◀ Agaves are a striking element of natural and cultivated landscapes in the Southwest.

Of all the threats facing the biosphere, from climate change and pollution to the spread of introduced species, none is more pressing than habitat loss and fragmentation. But as luck would have it, habitat fragmentation is the global challenge that the average gardener has the greatest ability to help solve. From massive landscape restoration projects to a few clay pots on the back patio, almost anyone can grow native plants and help knit together the fabric of ecosystems that have been torn apart by human activity.

As a tiny acorn can grow into a venerable old oak tree, what starts as a simple desire to beautify one garden may become a greater movement to restore watersheds, green our urban infrastructure, and improve quality of life for human and non-human beings alike by advocating for an equitable and sustainable balance between development and preservation. This book certainly doesn't have all the answers to the environmental challenges facing our region. However, it can serve as a jumping-off point for individuals, institutions, and communities to learn about the beautiful and resilient plants of the Southwest and how these species interact with some of the other living organisms—such as insects, birds, reptiles, and mammals—that we share our planet with.

The Southwest Native Plant Primer is based on the principles that:

- Native plants can be used to create aesthetically beautiful gardens.
- These landscapes can be lush and inviting without unsustainable water inputs.
- Gardens utilizing native plants will attract a wide variety of wildlife, and these visits are to be encouraged.
- Urban landscapes have a crucial role to play in larger efforts for the conservation of plants and animals.
- The more people who adopt these practices and beliefs, the greater the benefits will be to the constructed and natural ecosystems of the Southwest.

There are many ways to contribute to the cause of promoting native plant gardening: volunteering for habitat restoration efforts, joining your local native plant society, supporting local nurseries, or simply putting a seed in the ground and seeing what grows. Let us apply these principles to our landscapes and allow ourselves to be enchanted by the flora and fauna that have sustained human beings in this region for millennia.

◄ **A white-lined sphinx moth visits the blossoms of Hooker's evening primrose (*Oenothera elata*).**

Introduction

Writing a native plant gardening book to cover the American Southwest is a difficult task, not because the region is barren and devoid of life, but because the immense variety of our native flora presents gardeners with an embarrassment of riches. The Southwest is an astonishingly biodiverse region with habitats ranging from sun-scorched deserts to windswept tundra, with subtropical grasslands, dense chaparral, cool woodlands, and lush riparian canyons in between. This book is meant to inform readers about how to garden in the Southwest while imparting an appreciation for the myriad beautiful ecosystems found here.

◂ Bud prints contribute to the elegant aesthetic of this Parry's agave (*Agave parryi*).

This text covers a broad area, focusing on New Mexico, Arizona, and the southern portions of Nevada, Utah, and Colorado, into eastern California and southwestern Texas. There are vast differences in environmental conditions across this region, but there are also important commonalities. The plant list found in this book features some of the most prominent and commonly available species from each of our bioregions, as well as lesser-known plants that deserve a place in Southwestern gardens. Because there are so many potential plants to include, we have focused on selecting species that are either indicative of each bioregion or broadly distributed across multiple bioregions in the Southwest (to make this primer as useful for someone living in Durango, Colorado, as it is for someone in Yuma, Arizona). There are many plants available to gardeners that there was simply not room in this book to cover. See the resources section at the end and browse your local nurseries and botanical gardens to get additional ideas.

Increasingly, gardeners are becoming aware that their landscapes are part of the larger ecosystems they inhabit. For this reason, *The Southwest Native Plant Primer* will highlight the role of native plants in supporting pollinators, birds, and even reptiles, such as desert tortoises. This book will also delve into strategies for harvesting water for garden use, mitigating heat in our homes, and designing native plant gardens that mimic the natural communities found in the Southwest.

Our region hosts a unique assemblage of species that make for striking and resilient landscapes. The plant list in this book is organized by trees, shrubs, perennials, cacti and succulents, grasses, and vines. This primer will help people understand how to properly plant and care for these species and use them to maximum effect in their yards or public spaces. Gardening with native plants is about cultivating species that will flourish where they are planted to the benefit of garden and gardener alike. This is a book for those who want to work with the elements of the Southwest, not against them.

The rugged topography of the Southwest means that numerous habitats can exist within a small geographic area, as evidenced by the view from Mount Graham near the Arizona–New Mexico border.

The Southwest and Its Bioregions

The Southwest has it all. From Badwater basin in eastern California, more than 200 feet below sea level, to heights of over 13,000 feet in the peaks of northern New Mexico and southern Colorado. Average rainfall can range from 3 inches in some areas to nearly 40 inches in others, while temperatures can run from below freezing to well over a hundred degrees. The huge variation in climatic conditions around the Southwest makes it difficult to generalize about the region, but some basic axioms will be relevant to all gardeners.

◀ The Arizona Upland subdivision of the Sonoran Desert is particularly lush and inviting following abundant spring rains.

First off, the Southwest is arid compared to other regions of the United States. Even where rainfall is fairly high, average temperatures in the Southwest will tend to be higher than in other parts of the country at similar elevations. In addition, water is relatively scarce across most of the Southwest. Not only are surface water and groundwater precious resources, but high heat and an abundance of sunlight mean that precipitation from rainfall and snowmelt evaporates quickly out of the soil. Aridity is only increasing with the impacts of climate change, leading to higher temperatures and less predictable rainfall. Wherever you are in the Southwest, planning for heat and drought is essential.

Another interesting quirk about the Southwest's climate is its bimodal rainfall pattern. In other words, the Southwest has a winter–spring rainy season and a late summer–early fall rainy period known as the monsoon season. The impacts of this pattern are unevenly spread, with the amount of summer rain as a percentage of total annual

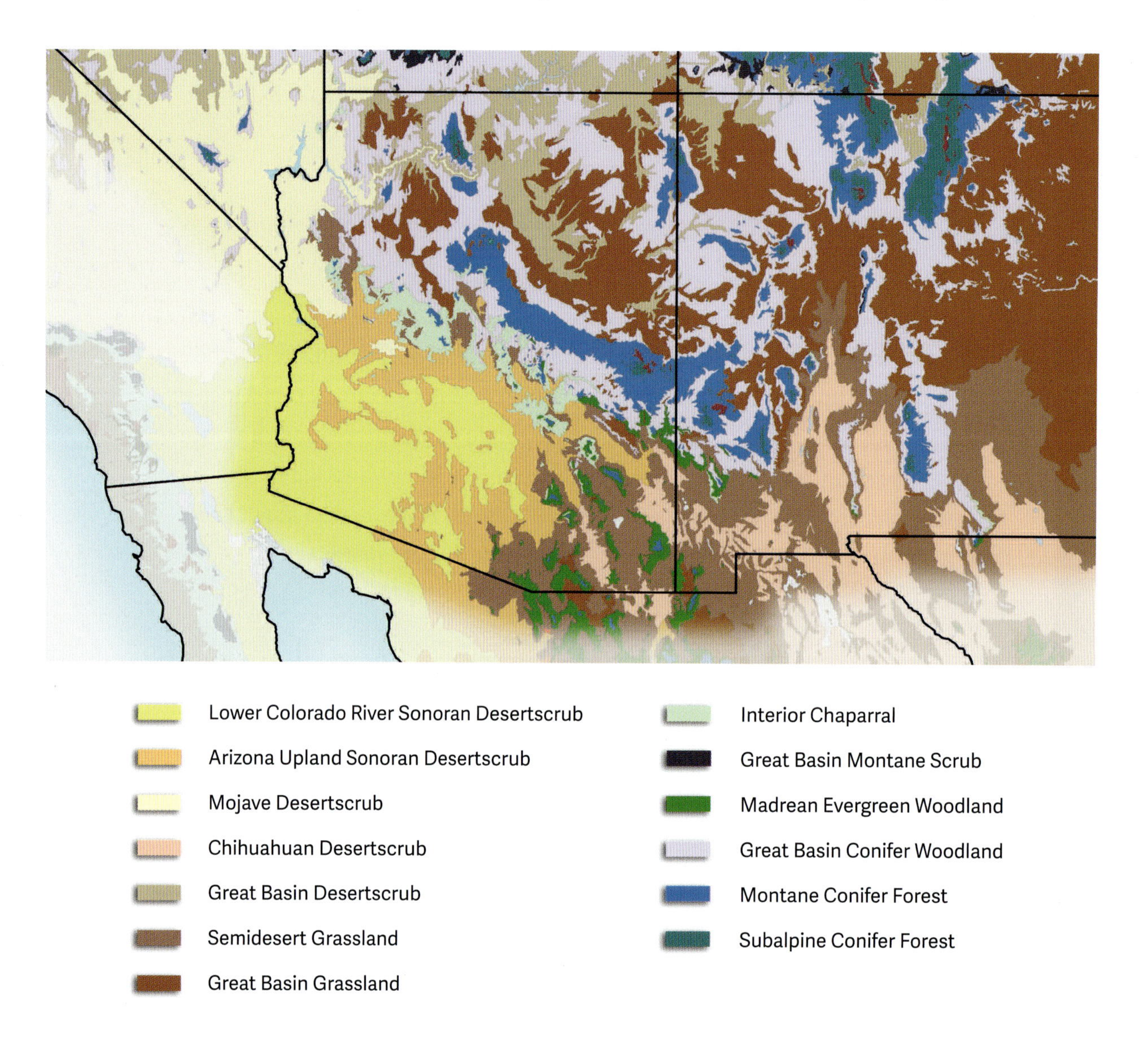

rainfall generally increasing as you move east. Las Vegas, Nevada, and St. George, Utah, will tend to have a higher percentage of winter rainfall, while Las Cruces, New Mexico, and El Paso, Texas, will tend to have a higher percentage of summer rainfall. Tucson, Arizona, on the other hand, will theoretically have about an even split of winter and summer precipitation. This pattern varies year-to-year and by location. The takeaway is that, in general, the Southwest has two rainy seasons, and this can be important for gardeners deciding when to install new plants and how much to water.

There are many systems to define the different regions of the Southwest, and each has its merits. Some systems are climatic, such as the US Department of Agriculture hardiness map. Others are geologic, like those put out by the US Geological Survey. Still others are based on plant communities. This book adapts the biotic community designations laid out in the pioneering work *Biotic Communities: Southwestern United States and Northwestern Mexico* (and the map titled *Biotic Communities of the Southwest*) created by David E. Brown and Charles H. Lowe. The biotic community system is ideal for gardeners because it is based on groups of indicative species and climatic conditions that can be transposed from natural to cultivated settings. It is important to remember that nature doesn't draw hard lines, and these biotic communities mix and mingle based on soil type, sun exposure, and hydrology. Understanding these biotic communities and how they relate to our cultivated landscapes will help you make appropriate plant selections that turn your yard from a lonely island into an active corridor that connects your garden to neighboring ecosystems.

Lower Colorado River Sonoran Desertscrub

This is one of the hottest and harshest environments in North America and is found at elevations ranging from just above sea level up to about 1500 feet. This biotic community is found in eastern California and southwestern Arizona, where it is defined by extreme aridity with average annual rainfall between 3 to 8 inches. Because of its dry climate and high temperatures, this region hosts some of the toughest plants on the continent, with creosote bush (*Larrea tridentata*) and white bursage (*Ambrosia dumosa*) dominating the dry flats, while most other trees and shrubs, such as honey mesquite (*Neltuma glandulosa*) and smoke tree (*Psorothamnus spinosus*), keep to the sandy washes where extra moisture can be found. Lower Colorado Sonoran Desertscrub also features extensive areas with very little perennial vegetation, which are covered in expanses of "desert pavement," a compact accumulation of rocks and sand that forms an almost impenetrable groundcover. Years with above-average winter rainfall may result in stunning displays of wildflowers that bloom and set seed in a riot of color before drying to a crisp with the onset of summer heat.

Creosote bush (*Larrea tridentata*) with a gall formed by a creosote gall midge (*Asphondylia* spp.).

Arizona Upland Sonoran Desertscrub

Around towns like Tucson, Ajo, and Wickenburg, Arizona, is a surprisingly verdant desert consisting of giant saguaro cacti (*Carnegiea gigantea*) and spiny trees like foothill paloverde (*Parkinsonia microphylla*), velvet mesquite (*Neltuma velutina*), and ironwood (*Olneya tesota*). These species are indicative of Arizona Upland Sonoran Desertscrub. This bioregion is distinctive for its high density of plant life supported by a bimodal rainfall pattern where roughly equal portions of the 8 to 13 inches of annual precipitation occur in winter and summer. Spread across an elevation range spanning 900 to 3500 feet, this richly vegetated desert hosts a wide variety of plant and animal life, including what may be the highest diversity of bee species found anywhere on the planet, a fact that makes pollinator gardening in this region particularly rewarding.

Despite its legendary heat, the Mojave Desert is home to a diversity of fascinating plant species.

Mojave Desertscrub

Found in parts of California, southern Nevada, Utah, and northwestern Arizona, the Mojave is the smallest of the American deserts but hosts a distinctive assemblage of plants, some of which are landscaping favorites. Spanning an elevational gradient from below sea level to around 3500 feet, the Mojave Desert is home to the iconic Joshua tree (*Yucca brevifolia*), one of the most famous of Southwestern species; beavertail pricklypear (*Opuntia basilaris*), a popular cactus for gardens; and abundant stands of flattop buckwheat (*Eriogonum fasciculatum*), one of the Southwest's best pollinator plants. Mojave Desert habitats with annual rainfall ranging from 4 to 9 inches can be found around Twentynine Palms, California; Las Vegas, Nevada; St. George, Utah; and west of Kingman, Arizona.

Chihuahuan Desertscrub

Unlike the Arizona Upland Sonoran Desert, with its roughly equal rainy seasons, the 10 to 14 inches of annual rainfall in the Chihuahuan Desert falls mostly in summer in the form of intense and sometimes violent monsoon rains. The Chihuahuan Desert is massive in size, stretching well into northern Mexico and western Texas and covering almost all of southern New Mexico, including the cities of Deming and Socorro, with the western boundary reaching southeastern Arizona around the historic town of Tombstone. Many Chihuahuan Desert species are found nowhere else in North America, but some of the more common plants include creosote bush (*Larrea tridentata*), littleleaf sumac (*Rhus microphylla*), mariola (*Parthenium incanum*), and ocotillo (*Fouquieria splendens*). The Chihuahuan Desert is known for its wild climatic swings and cooler temperatures than the Sonoran or Mojave deserts. Because of this, Chihuahuan species tend to be relatively cold-tolerant.

Ocotillo (*Fouquieria splendens*) is one of many species that thrive in the large expanse of the binational Chihuahuan Desert.

Great Basin Desertscrub

Cold but dry, Great Basin Desertscrub is a challenging environment where plant communities are shaped by the freezing winter temperatures and the 7 to 12 inches of rainfall they can expect in an average year. This is the land of big sagebrush (*Artemisia tridentata*), which covers large expanses of this biotic community, along with other hardy shrubs and succulents like winterfat (*Krascheninnikovia lanata*), plains pricklypear (*Opuntia polyacantha*), and narrowleaf yucca (*Yucca angustissima*). Found between 3900 and 8500 feet, representative Great Basin Desertscrub communities can be seen around Taos, New Mexico; Bluff, Utah; Cortez, Colorado; and Page, Arizona.

Great Basin Desertscrub reaches its southernmost extent in the northern portions of Arizona and New Mexico.

Semidesert Grassland

Alternatively called Scrub Grassland, Apacherian Savannah, and Short Grassland, this habitat is distinguished by extensive plains of warm-season grasses such as sideoats grama (*Bouteloua curtipendula*), bullgrass (*Muhlenbergia emersleyi*), and tanglehead (*Heteropogon contortus*). Primarily found in the US–Mexico borderlands between about 3500 and 6000 feet, this habitat can be seen around towns like Patagonia, Arizona, and Lordsburg, New Mexico, with patches as far northwest as Kingman, Arizona. At its lower elevations, Semidesert Grassland generally borders Desertscrub communities, and at its upper reaches, it runs into Interior Chaparral and Madrean Evergreen Woodland. These grasslands have changed substantially over the last few centuries with the introduction of cattle and other livestock and the advent of fire-suppression efforts that have caused an increase in the number of woody shrubs and succulents like velvet mesquite (*Neltuma velutina*), soaptree yucca (*Yucca elata*), and ocotillo (*Fouquieria splendens*). This community receives between 10 and 17 inches of rainfall annually, much of it in summer when the dry, tan hillsides explode into a riotous profusion of green grasses and colorful wildflowers.

The wide-open spaces of Semidesert Grassland offer some of the best views in the Southwest.

Great Basin Grassland

Colder, wetter, and farther north than Semidesert Grassland, Great Basin Grassland features a distinctive assemblage of short and tall grasses, including little bluestem (*Schizachyrium scoparium*), slender wheatgrass (*Elymus trachycaulus*), and Indian ricegrass (*Achnatherum hymenoides*). Like Semidesert Grassland, the species demographics of this biotic community have been altered by human activity and the increasing spread of shrubs, succulents, and trees such as junipers (*Juniperus* spp.) and fourwing saltbush (*Atriplex canescens*). This bioregion typically receives between 11 and 20 inches of rain per year, and these rains can bring out stunning displays of wildflowers such as tufted evening primrose (*Oenothera cespitosa*) and globemallows (*Sphaeralcea* spp.). You can find this biotic community between about 4000 and 7500 feet around Albuquerque, New Mexico, and Holbrook, Arizona, with isolated patches as far south as the San Rafael Valley in Arizona.

Great Basin Grassland hosts a blend of grasses, wildflowers, and shrubs that are tolerant of cold temperatures and full-sun exposure.

Interior Chaparral

Across a narrow swath of central Arizona, between Desertscrub and Woodland habitats and at elevations from 3300 to 6600 feet, you can find stretches of densely vegetated Interior Chaparral. This fire-prone biotic community is dominated by a diverse assemblage of woody shrubs such as pointleaf manzanita (*Arctostaphylos pungens*), scrub oak (*Quercus turbinella*), cliffrose (*Purshia stansburyana*), and sugar sumac (*Rhus ovata*), many of which can also be found in California chaparral communities. With average rainfall between 13 and 25 inches per year, this is a community rich in flora and fauna that can be appreciated around Bagdad and Miami, Arizona, and the Sierra Ancha and Mazatzal mountains.

Great Basin Montane Scrub

In the mountain foothills surrounding Durango, Colorado; Mesa Verde National Park; and Bryce Canyon National Park in Utah between elevations of 7500 to 9000 feet, you may find yourself in the chaparral-like Great Basin Montane Scrub community. Rainfall is relatively low, averaging between 14 and 21 inches annually, and shrubs and small trees like Gambel oak (*Quercus gambelii*), mountain mahogany (*Cercocarpus montanus*), Fendler's ceanothus (*Ceanothus fendleri*), Woods' rose (*Rosa woodsii*), and New Mexico locust (*Robinia neomexicana*) form stands that buffer the higher Montane Conifer Forest from the lower Great Basin Desertscrub community.

Pointleaf manzanita (*Arctostaphylos pungens*) is just one of many shrubs that form nearly impenetrable stands in Interior Chaparral habitats.

Though limited in geographic extent, Great Basin Montane Scrub forms an important buffer zone between desert and forest habitats.

Madrean Evergreen Woodland

The mountains that straddle the boundaries of Arizona, New Mexico, and the Mexican state of Sonora are often called "sky islands" for the way they form isolated patches of woodland and forest surrounded by a sea of desert. Madrean Evergreen Woodland is an extremely biodiverse habitat that covers the slopes above Semidesert Grassland and Desertscrub communities, providing a haven for wildlife that can't survive in the harsher climates below. This woodland is notable for its strong links to the more subtropical Sierra Madre Occidental mountain range of northern Mexico, sharing many species, including alligator juniper (*Juniperus deppeana*), Mexican blue oak (*Quercus oblongifolia*), and kidneywood (*Eysenhardtia orthocarpa*). Madrean Evergreen Woodland forms an important ecological tidepool between the Rocky Mountains of Canada and the United States, the Sierra Madre of Mexico, and the Sonoran and Chihuahuan deserts. With average rainfall from 15 to more than 30 inches per year, this biotic community can be found at elevations of 4000 to 7500 feet. Despite occurring at higher elevations, this habitat supports a rich assemblage of succulents, including Palmer's agave (*Agave palmeri*), beargrass (*Nolina microcarpa*), and mountain yucca (*Yucca schottii*). Prime examples of Madrean Evergreen Woodland can be found in the Santa Rita, Chiricahua, Pinaleño, and Animas mountains of Arizona and New Mexico.

Madrean Evergreen Woodlands host species with subtropical affinities, such as Mexican blue oak (*Quercus oblongifolia*).

Great Basin Conifer Woodland

Above Great Basin Desertscrub and Grassland and below Montane Conifer Forest communities, extensive patches of Great Basin Conifer Woodland cover large portions of the central and northern Southwest. This biotic community, with rainfall averaging 10 to 20 inches per year, is best distinguished by the dominance of pinyon pine (*Pinus edulis*) and various species of juniper (*Juniperus* spp.). These woodlands experience extremely cold temperatures in wintertime and have a relatively brief growing season, so the trees found here are short and dense, often more shrub- than tree-like. Between the canopies of these coniferous species, other plants such as ricegrass (*Achnatherum hymenoides*), blue grama (*Bouteloua gracilis*), and winterfat (*Krascheninnikovia lanata*) cover the thin, rocky soil. Look for Great Basin Conifer Woodland between 4000 and 7500 feet around Santa Fe and Silver City, New Mexico; Sedona, Arizona; and Zion National Park in Utah.

Great Basin Conifer Woodland skirts the bases of the red mesas and spires of Sedona, Arizona.

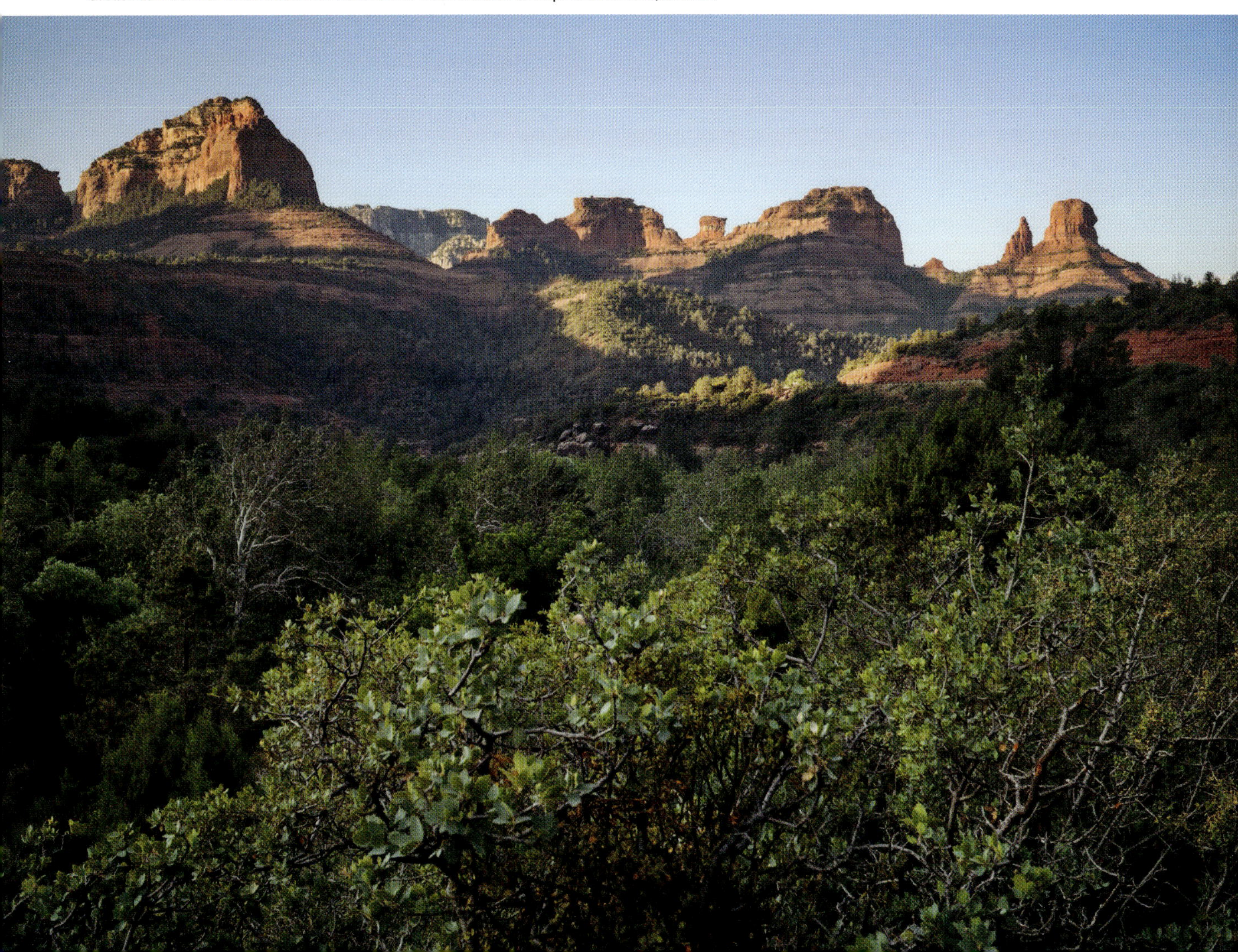

Montane Conifer Forest

This widespread biotic community is primarily found between 7000 and 9000 feet, where it may occur as isolated patches on top of mountains or as expansive forests in the northern reaches of the Southwest. With annual rainfall between 18 and 30 inches, these semi-arid forests feature stands of ponderosa pine (*Pinus ponderosa*) with an understory consisting of cold-tolerant species such as Gambel oak (*Quercus gambelii*), pine dropseed (*Blepharoneuron tricholepis*), Fendler's ceanothus (*Ceanothus fendleri*), creeping barberry (*Mahonia repens*), and currants (*Ribes* spp.). At higher elevations, this assemblage involves other trees, such as quaking aspen (*Populus tremuloides*) and bigtooth maple (*Acer grandidentatum*). Representative stands of Montane Conifer Forest can be seen in the Santa Catalina, Pinaleño, and Sacramento mountains in the southernmost part of the Southwest, as well as in Flagstaff and Pinetop–Lakeside, Arizona, and north of Taos, New Mexico.

Pines (*Pinus* spp.) are an essential element of the Montane Conifer forests of the Southwest.

Subalpine Conifer Forest

At the highest reaches of mountain ranges in the Southwest (above 8000 and up to more than 12,000 feet), Subalpine Conifer Forest bridges the gap between Montane Conifer Forest and a timberline above which no large shrub or tree species grow. Rainfall here is high for the Southwest, averaging 25 to 40 inches per year, an amount that supports dense forests of blue spruce (*Picea pungens*), white fir (*Abies concolor*), and

Subalpine Conifer Forest also hosts deciduous trees such as maples (*Acer* spp.), which offer a fall display that rivals the spring wildflower shows of the low desert.

quaking aspen (*Populus tremuloides*), with an understory of common juniper (*Juniperus communis*), shrubby cinquefoil (*Potentilla fruticosa*), and various currants (*Ribes* spp.). The cold tolerance of these species makes them suitable for high-elevation gardens, but their intolerance of heat and sensitivity to drought means they should be used sparingly and placed in appropriate microclimates where they can thrive. This biotic community can be found in the San Francisco Peaks of northern Arizona and the Sangre de Cristo and San Juan mountains that span southern Colorado and northern New Mexico, with small outposts found even farther south on Mount Graham in southern Arizona.

Riparian Corridors

Cutting across all elevations and biotic communities are veins of seasonal or perennial water called Riparian Corridors. At lower elevations, this community might occur in seasonally dry washes and feature catclaw acacia (*Senegalia greggii*) and desert hackberry (*Celtis pallida*), with velvet ash (*Fraxinus velutina*), netleaf hackberry (*Celtis reticulata*), and Fremont's cottonwood (*Populus fremontii*) becoming common as you move up in elevation. Higher still, Arizona sycamore (*Platanus wrightii*) and cypress (*Hesperocyparis arizonica*) fill the canyons before even these are replaced by quaking aspen (*Populus tremuloides*) and boxelder maple (*Acer negundo*). Riparian Corridors may be found along the rivers of the Southwest or in the canyons that punctuate the many mountain ranges of the region. Wherever they occur, they are sure to host a high level of biodiversity. These habitats are some of the most threatened in the Southwest and some of the most stunningly beautiful. Riparian plants are not always suitable for gardens, as they invariably require high water inputs and will be more sensitive to extreme heat or a missed irrigation.

Riparian Corridors are the most biodiverse habitats in the Southwest, despite covering only a tiny fraction of the region's total area.

What Is a Native Plant?

It's worth beginning by asking ourselves: What is a native plant? The answer is ... it depends. The definition of "native" is something that will vary from gardener to gardener, depending on what it is they want to accomplish with their landscape. In other words, if your goal is to restore the habitat that may have been present prior to the construction of your home, you might define native plants as those that occur naturally near your garden in a 5- to 10-mile radius. If your focus is to attract a particular type of wildlife—let's say you want birds in your yard—you might draw from bird-attracting species that occur within 50, 100, or 500 miles of your home in ecosystems with similar environmental conditions.

◀ Saguaro (*Carnegiea gigantea*) is one of the Southwest's most iconic species.

Because borders rarely adhere to ecological reality, thinking about native plants through the lens of political boundaries isn't necessarily helpful. For instance, a plant that occurs in the deserts of northern Sonora in Mexico may be more native to a Tucson, Arizona, garden than a plant that occurs in the pine forests around Flagstaff, Arizona. This is because the Sonoran Desert spans the United States and Mexico, acting as a biological thread between these two countries. Additionally, thinking about native species as those that occur within a certain number of miles from your home may be misleading because of the incredibly varied topography of the Southwest. With mountains abruptly rising thousands of feet from valley floors, the parched deserts and temperate subalpine forests of the Southwest may be found within just a few miles of each other as the crow flies. A blue spruce (*Picea pungens*) will quickly wither in a desert garden just as a saguaro (*Carnegiea gigantea*) will freeze in a montane landscape.

A chimney bee (*Diadasia* spp.) forages for food in the flower of a pencil cholla (*Cylindropuntia arbuscula*).

For these reasons and more, it is good to take a bioregional approach to selecting plants for your landscape. In deciding how you will define *native*, there is no substitute for getting out of your yard and taking a hike. Go somewhere nearby at a similar elevation to your home and see what is growing there. Take note of what species you see on clay flats, rocky slopes, and sandy washes, and then assess your soil and topography at home using the techniques found later in this book. Spend time talking to your local nursery staff, visit nearby public gardens, and take a stroll around your neighborhood. See if you recognize the same plants you have found on hikes, and most of all, experiment. Losing plants can be a frustrating and potentially expensive way to learn, but you will quickly become familiar with best practices for your landscape. Remember that not every seed grows to become a mature tree in nature either. Evolution is one big exercise in trial and error, and so is gardening. By taking a bioregional approach to plant selection, you can minimize losses and build your habitat more quickly by starting with the right plant for the right location.

Why Use Native Plants?

You may be asking whether all this talk about native plants really matters. Does it make a difference if I plant a native versus an introduced species? Will the birds and bugs actually notice? The answer is unequivocally *yes*. Native plants have evolved within their respective ecosystems over the course of millennia, becoming ideally suited to withstand the particular pressures and conditions of temperature, rainfall, and soil within their distinct habitats. But they have not evolved alone. A variety of organisms ranging from fungi and bacteria to reptiles, birds, and mammals have evolved to use the food and shelter provided by the plants that grow near them. Gila woodpeckers (*Melanerpes uropygialis*) are well-adapted to carving out

A garden full of native plants can be lush and full with only minimal water inputs, as seen in the front yard of this Tucson home.

nests in the trunks of saguaros (*Carnegiea gigantea*). rufous hummingbirds (*Selasphorus rufus*) perfectly time their long migrations to the availability of certain nectar plants, such as ocotillo (*Fouquieria splendens*). And yucca moths (*Tegeticula* spp.) mature into adults just as yucca flowers, into which the moths will insert their eggs, begin to appear.

These are just a few of the impossibly complex relationships that exist between the living beings of the Southwest. Our lack of understanding or outright ignorance of these types of relationships is all the more reason not to disturb them. Planting the correct native species for your location directly supports local wildlife and allows for ease of cultivation that is not attainable with many non-native plants.

The good news is that the Southwest provides no shortage of plants to choose from. The species covered in this book are just the tip of the iceberg when it comes to native plants available to gardeners in our region. With thousands of native plants—many hundreds of which can be found as seeds or starts from local growers—every shape, size, texture, and color of plant that any gardener could need can be found right here among the diverse and distinctive flora of the Southwest. As you begin your journey into the world of native plant gardening, you are more likely to be

A hawkmoth (*Manduca* spp.) rests on a pricklypear pad.

overwhelmed by the wealth of options than frustrated by the lack of choices for your landscape.

A yard full of non-native plants can be visually beautiful, it's true. But this beauty often serves to obscure what is, ultimately, a lovely wasteland. With that said, a garden populated by native plants can also play host to a few species from elsewhere. Perhaps, like many gardeners, you want to grow fruits and vegetables to provide nutritious and tasty food for your loved ones. The water saved by using native species will ensure that needier fruit and vegetable cultivars can be grown without breaking the bank, while the pollinators attracted by native plants will likely find their way to your squash and tomatoes—particularly if you cultivate heirloom varieties with a long history in your area. You may also have an affinity for potted tropical plants, showy annual bedding flowers, or a species from back home that you feel you simply can't live without. That is OK; you don't need to be dogmatic.

However, it is essential that you do your homework and avoid plants that may become invasive. Some of the most pernicious weeds in the Southwest, from crimson fountaingrass (*Pennisetum setaceum*) to tamarisk (*Tamarix* spp.), were intentionally introduced as landscaping species and have since "hopped the fence" and displaced or eliminated native plants, pollinators, and wildlife from their former homes. And who knows? With time, you may find that you lose interest in those non-native species you thought you needed as you see the myriad benefits native plants impart to your landscape.

Native Alternatives to Non-native Species

Even as the popularity of native plant gardening continues to grow, there are still nurseries and landscapers who cling to a handful of introduced species as if they are the only plants on Earth. A perusal of your local garden center, or a drive through any suburban neighborhood, will turn

An Apache jumping spider (*Phidippus apacheanus*) dines on a grasshopper that it caught from the blade of a native grass.

INTRODUCED SPECIES	NATIVE ALTERNATIVES
Afghan and Aleppo pine (*Pinus brutia* var. *eldarica*, *P. halepensis*)	pinyon or ponderosa pine (*Pinus edulis*, *P. ponderosa*)
African sumac (*Searsia lancea*)	desert willow (*Chilopsis linearis*)
Argentine saguaro (*Leucostele terscheckii*)	saguaro (*Carnegiea gigantea*)
Bermuda grass (*Cynodon dactylon*)	curly mesquite grass (*Hilaria belangeri*), vine mesquite grass (*Hopia obtusa*)
Chinese elm (*Ulmus parvifolia*)	netleaf hackberry (*Celtis reticulata*)
crimson fountaingrass (*Pennisetum setaceum*)	bullgrass (*Muhlenbergia emersleyi*)
heavenly bamboo (*Nandina domestica*)	red barberry (*Mahonia haematocarpa*)
Italian cypress (*Cupressus sempervirens*)	Arizona cypress (*Hesperocyparis arizonica*)
lantana (*Lantana camara*)	Goodding's mock vervain, Dakota mock vervain (*Glandularia gooddingii*, *G. bipinnatifida*)
Moroccan mound (*Euphorbia resinifera*)	claret cup hedgehog cactus (*Echinocereus coccineus*)
oleander (*Nerium oleander*)	Arizona rosewood (*Vauquelinia californica*)
oriental arborvitae (*Thuja* spp.)	Rocky Mountain juniper (*Juniperus scopulorum*)
pink muhly (*Muhlenbergia capillaris*)	plains lovegrass (*Eragrostis intermedia*)
red push pistache (*Pistacia* × 'red push')	velvet ash (*Fraxinus velutina*)
southern and Texas live oaks (*Quercus virginiana*, *Q. fusiformis*)	Emory oak (*Quercus emoryi*)
wax-leaf privet (*Ligustrum japonicum*)	sugar sumac (*Rhus ovata*)
yellow bird of paradise (*Erythrostemon gilliesii*)	prairie acacia (*Acaciella angustissima*)

Gardening with native plants connects your landscape to the wider ecological webs that support life in the Southwest.

up a plethora of introduced species that ought to have disappeared from our landscapes long ago. As much as the dedicated native plant gardener might be tempted to sneak out at night and fell their neighbor's African sumac (*Searsia lancea*) or yank their heavenly bamboo (*Nandina domestica*) out by the roots, it is much better for mental health and domestic bliss to lead by example. Even the most stubborn snowbird or intransigent HOA board may be won over to the cause of native plants through a quick fence-side conversation or an invitation to tour a yard populated with thriving native specimens.

Nurseries can also be made to change, as growers will always want to produce what sells. If customer after customer comes in looking for native landscaping plants, and if consumers refuse to buy the same old introduced plants these nurseries have been churning out for decades, they will get the message. Eventually, they will alter their practices or risk going out of business with greenhouses packed to the rafters with non-native plants that nobody wants anymore. A list of some native alternatives to commonly planted introduced species can be found on the proceeding page; it can serve as a jumping-off point as you scour local nurseries and rummage through seed catalogs to design your landscape.

Creating Biodiversity in Our Backyards

This is the point where, in other plant books, you'd come to a section titled "Pest Management," where you would learn about the best techniques for repelling critters and annihilating bugs to prevent anything from touching so much as a leaf, bud, or flower in your well-manicured landscape. But you aren't reading that book. It is important to keep in mind that our urban landscapes do not exist in a vacuum. Each gardener can create a small habitat that connects them to the ecosystems found around their communities, towns, and regions. Your yard can support a diverse array of pollinators and other insects, serve as a way station for local and migrating birds, and be a buffet for reptiles such as desert tortoises. As the seasons change, different pollinators and migrating birds will arrive in your garden before moving on, and as your garden grows, each year will bring a different dynamic and new visitors. Real, practical conservation can occur at a backyard scale, not just through proper plant selection but also through sustainable practices such as water harvesting, composting, and gardening without chemicals. In this way, a landscape can serve as a micro-refuge for wildlife. As gardening with native species continues to catch on, whole communities can come together to play a role in protecting the biodiversity of the Southwest.

A bordered mantis (*Stagmomantis limbata*) waits to capture unsuspecting prey.

A moth mimics the color and shape of a net-winged beetle (*Lycid* spp.), showcasing the amazing powers of deception possessed by our native insects.

Tent moth caterpillars (*Malacosoma* spp.) dine on the leaves of Gambel oak (*Quercus gambelii*). These caterpillars will come and go quickly, doing no permanent damage to the plant.

The responsibility to use our gardens for good requires that we not only accept that wildlife will come into our yards, but that we do our best to encourage it. Luckily, attracting and supporting wildlife is less about the work we must do and more about the tasks we can dispense with. For example, many common birds eat seeds, so simply choosing not to deadhead your plants immediately after they flower will preserve an important resource for avian fauna and save you money on bagged birdseed.

One of the most basic needs for any organism is shelter. When we remove every leaf and twig that falls in our landscape, put them into trash bags, and send them to the landfill, we not only lose organic matter for our soils, but we reduce the potential habitat around our homes for everything from tiny encrusting termites (*Gnathamitermes perplexus*) to cottontail rabbits (*Sylvilagus* spp.) and Gambel's quail (*Callipepla gambelii*). Creating a brush pile in an out-of-the-way corner of the yard will save you a trip to the dump and improve the overall wildlife value of your landscape.

A verdin (*Auriparus flaviceps*) rests on the branches of a creosote bush (*Larrea tridentata*).

Tortoises, such as this Sonoran desert tortoise (*Gopherus morafkai*), are fantastic garden pets that can graze your native plants. This tortoise is particularly fond of Engelmann pricklypear (*Opuntia engelmannii*) fruits.

Gardening for reptiles is becoming increasingly popular, especially in Arizona, where the state Game and Fish Department actively seeks homes for captive-raised or rescued desert tortoises (*Gopherus morafkai* and *G. agassizii*). Tortoises make surprisingly charming and sociable pets, and they do best when they have a diverse array of native plants available for them to eat and shelter under. While many nurseries and public gardens have lists of food plants suitable for tortoises, they will also be indicated in the plant profiles found within this book.

Butterflies and their more enigmatic cousins, moths, lay their eggs on host plants that will serve as a food source for their caterpillars when they emerge. Almost every single native plant in the Southwest serves as a larval food source for some butterfly or moth, so planting a wide selection of species and including multiples of each will bring generation after generation to your yard. After exiting the chrysalis, these insects will reward you by gracing your yard with their colorful patterns and valuable pollinator services.

The southwestern United States is a global hotspot for bee diversity, and most native species are solitary, stingless, and very effective pollinators of native plants. Some of these bee species make their homes in decaying wood, so leaving an old agave stalk or a dead tree limb in place will meet their needs. Many bees are ground-dwelling, and when you leave loose dirt or fallen plant matter on the ground instead of raking it up or scattering it with a blower, you provide a haven for these species. Others prefer to make their homes in dense shrubs or tufts of grass, so abstaining from unnecessary trimming (which often diminishes the aesthetic value of the plant anyway) will help make your landscape into a pollinator nursery. As you can see, one of the best ways to make your garden a wildlife haven is to eliminate extraneous tasks from your to-do list and spend the time you saved listening to the birds chirp and the pollinators buzz.

To encourage butterflies such as this queen (*Danaus gilippus*), you need to provide the larval food source, in this case milkweeds (*Asclepias* spp.); safe sites for the chrysalis, such as this ocotillo (*Fouquieria splendens*); and food for the adult butterfly when it emerges.

Water attracts this Anna's hummingbird (*Calypte anna*) to a garden. Providing sources of moisture will ensure your yard is a hotspot for all types of wildlife.

Wasps and bees flock to an aquarium tank filled with native insects and fish as well as yerba mansa (*Anemopsis californica*).

Keep in mind that the limiting factor for most life in the Southwest is the availability of water. Providing water in the form of a bird bath, a water-recycling pond, or even an aquaponics tank will bring in a host of species that would otherwise be absent, from birds to dragonflies. Even a patch of mud on the ground can entice wildlife. Male butterflies often mass around wet spots in a behavior known as "puddling," where they drink up the mineral-infused liquid for sustenance and then pass it on to potential mates in what entomologists call "nuptial gifts."

Bee-, bird-, and bat houses are increasingly popular, requiring nothing more than some wood, screws, and different-size drill bits; these easy-to-make structures can be the difference between your yard being a quick stop or a permanent residence for local wildlife.

Anybody who has ever turned compost can tell you that these piles of detritus are practically writhing with insect activity. Dig a pit in your yard or build a simple bin out of

Curve-billed thrashers (*Toxostoma curvirostre*) with bread taken out of a backyard compost pile.

Fluffgrass (*Dasyochloa pulchella*) and trailing windmills (*Allionia incarnata*) are native plants that are often pulled or sprayed with herbicide because they are considered "weeds." Plants like these should be considered a blessing, not a curse.

old pallets and fill it with food scraps and yard trimmings. Insects will appear, as if out of thin air, to help break down the material, and the birds and mammals that eat these insects will soon follow. Not only will composting increase your backyard biodiversity, but it will also improve your soil quality as you put nutrient-rich black gold from your pile back into your garden.

Aside from these techniques, the very best way to attract a diversity of wildlife to your garden is to provide a diversity of food sources. Every new plant you add to your landscape is a potential source of food or shelter for another insect, bird, reptile, or mammal. A good garden design considers not only the aesthetic traits of each plant but also what it has to offer as a source of backyard biodiversity.

Attracting an array of insects, birds, and mammals is in and of itself the best pest control. By creating a diverse habitat, you can help prevent any one species from taking over, maintaining the equilibrium that has allowed our natural landscapes to grow and thrive, insects and all, for millennia.

Weeds

Any garden will have its share of weeds. Weeds are just plants that tend to thrive in disturbed areas and fill in empty spaces where they will quickly grow and reproduce. A weed can be a native or a non-native plant, so some weeds may be benign or beneficial. In fact, some of the plants described in this book are often pulled or sprayed as "weeds" by landscapers or homeowners keen to eliminate anything unfamiliar from their yards. Other weeds may be seriously invasive plants that should be removed right away before they spread into adjacent yards or natural areas. The key is to get to know the plants germinating in your yard before you remove them so that you don't eliminate a plant you will later go to a nursery and purchase. Many agricultural extension offices or native plant societies will have an info line or email address you can contact to send photos and get identification help.

Getting Started

Assessing the Particulars of a Site

Before you ever put a shovel into the soil, it is important to take the time to assess the planting site for a variety of factors that will determine where plants should be located and how likely they are to thrive. By considering sun exposure, soil quality, temperature extremes, and water availability before planting, you will set your plants up for success and guarantee that they will not only survive but thrive.

◀ With defined paths, good plant selection, and basins to capture rain runoff, this in-progress garden is being set up for long-term success.

A gardener in training takes stock of the plants in a botanical garden. Observing what grows well in nearby public and private gardens will help you determine which species are right for you.

Sun Exposure

A rite of passage for gardeners in the Southwest is purchasing a plant labeled "full sun" and seeing it wither away in the blazing light of June. In a region like the Southwest, the intense sunlight and heat can be lethal to plants if they are not well adapted. However, structures and large trees or shrubs will create different exposures that can be used to the benefit of the gardener. This makes it essential to plan for the mature size of trees and shrubs because they

These ponderosa pines are thriving in full sun at an elevation of 7500 feet. These same trees would suffer in similar conditions at 1500 feet. Exposure and elevation are essential factors in determining what species will do well in your garden.

can significantly alter light exposures in a yard over time, which can be a positive or negative depending on the situation. Considering ways to utilize different exposures can allow a gardener to expand the range from which they draw plants. For instance, a cold-sensitive cactus such as a saguaro (*Carnegiea gigantea*) may be grown at higher elevations if planted near a hot, south-facing wall.

It is important to take note of the changes in the angle of the sun throughout the year. In summer, a plant on the northwest side of a structure may get blazing afternoon sun but be in shade all day in winter. The sun appears higher in summer and will hit your yard differently than in winter when the sun angle is much lower. Ironically, sunburn on cacti and succulents can be more prevalent in winter than in summer because trees may be defoliated, and the sun is shining on a large part of the plant's surface that is less well-protected than the growth tip. Sun angle and shadow calculators are easily found online, but there is no substitute for observation over the course of a year.

Goodding's mock vervain (*Glandularia gooddingii*) prefers well-drained sandy or rocky soils and may languish in heavy clay.

Soil

The Southwest offers a variety of soils, from barren alkaline clays to acidic forest loams and from scorching dunes to boggy meadows. Soil type and quality can differ wildly, even across a single garden, a fact you can leverage to create distinctive zones within a landscape.

At the most basic level, soils are graded along a scale from acidic (low pH) to alkaline (high pH) and from clay (made of tiny particles that hold moisture and are easily compacted) to sand (made of larger particles that quickly dry out and are relatively loose). In between these extremes are a variety of soil types, and assessing where your soil falls on this spectrum will help you determine what native plants are most appropriate for your conditions. An expensive full-spectrum soil analysis is unnecessary for the average gardener, and you can get a good idea of the qualities of your soil through a couple of simple tests.

pH

To test pH, you can buy a test strip. They are cheap and easy to use, only requiring that you mix soil with distilled water and insert a strip. However, another option is to use items that are available in most pantries. You can start by placing a couple tablespoons of soil into a glass of distilled water and adding half a cup of baking soda. If the mixture fizzes, it indicates that you have an acidic soil. You can repeat the test but add vinegar instead of baking soda, and if that solution fizzes, then your soil is alkaline. If neither mixture causes fizzing, then your soil has a neutral pH. This method is not particularly precise but gives you a general sense of the pH of your garden soil.

There is an overwhelming array of soil amendments, nutrients, and conditioners on the market today, and it can be tempting to want to micromanage your soil to push growth and develop neutral soil, one right in the middle of the pH scale, with pH 7. The important thing to keep in mind is that native plants of the Southwest are used to what would generally be considered "poor" soils with relatively high alkalinity and low availability of nutrients like phosphorus, iron, and zinc. These are the soils most of our native plants have adapted to over the course of millennia. The takeaway is that while our soils may not be ideal for growing a prize pumpkin or a happy rose or rhododendron, they are already well suited to native plants. Rather than going out and buying expensive and possibly counterproductive amendments, you can select native plants that are naturally accustomed to your soil type.

As its common name suggests, alkali sacaton (*Sporobolus airoides*) prefers soils higher on the pH scale and is tolerant of salty substrates.

TEXTURE

There are two easy ways to test soil texture. One is to fill a jar about a third of the way with sifted soil and then top off the jar with water. Shake the jar and allow the contents to settle over the course of 24 to 48 hours. Sand will settle on the bottom, silt on top of that, and clay will form the top layer. By taking a ruler and measuring the respective height of each distinct horizon, you can get a good sense of the type of soil you are working with. An even simpler test is to dig out a handful of soil from your garden, wet it thoroughly, and squeeze it in one hand. If the soil crumbles, it is sandy, but if it stays together and forms a play dough–like ball, then it is clay. By repeating this test in different parts of your yard, you can quickly and effectively get a sense for the

Looking out across the Sonoran Desert as the sun sets on a spring day.

soil you are working with and can either amend your soil or base your plantings on species tolerant of the conditions you already have.

At the extreme ends of the spectrum, you may have soils that are almost entirely sand, a situation that sometimes occurs in houses built on floodplains or at the bases of eroding peaks. Sandy soils will mean that water drains quickly, so little moisture is retained. These types of soils favor succulents that prefer good drainage and can rot in wet soils. Soils that are hard-packed clay are more suitable for shallow-rooted grasses and perennials that are less susceptible to rot and better able to work their roots into small crevices and openings in the clay, loosening and oxygenating it over time.

One long-term solution for balancing both types of extremes is the addition of compost and mulch. Compost can be purchased, or even better, made on-site from the food scraps, yard trimmings, and cardboard that are the inevitable byproducts of any household. Over time, compost will tend to balance the pH of soils, create better texture in sandy situations, and contribute greater oxygen exchange and fertility to clay soils. It is important to remember that you can have too much of a good thing and that the addition of too much compost can create a situation that is actually unfavorable to some native plants. Additionally, digging a hole and simply dumping in the compost in place of native soil can induce a "container effect," where the roots of your plant stay within the planting hole and don't expand outward, so blending your native soil with the compost is essential.

CALICHE

Caliche is a concrete-like layer of calcium carbonate that builds up because of a lack of rain to flush calcium from the soil. Caliche can create an impermeable layer that prevents proper drainage and makes important micronutrients

Oak tree canopies create a microclimate that is cooler in summer and insulated in winter.

unavailable to your plants. While composting, mulching, and the action of plant roots can help break up caliche over time, in the short term, you may need to take drastic action. The most common solution is to dig drainage holes through the caliche layer so that water can pass through. Depending on the thickness of the caliche layer, this may involve the use of a metal digging bar or a jackhammer. Loose caliche should be removed from the hole. More extreme treatments involving the use of acids to dissolve the caliche are likely unnecessary for native plants, and the use of tools should be sufficient to break up the caliche layer enough to allow for planting.

Heat or Cold

The Southwest is notorious for its swings in temperature. This is true across the region, though the extremes differ depending on what bioregion a gardener is in. Regardless of whether heat or cold is your primary issue, the solutions are similar.

To begin with, it is well documented that the canopy of a mature tree can significantly reduce soil temperatures in summer, while the shade has the added effect of helping keep moisture in the ground. This cooling effect is especially pronounced in urban areas where increasing amounts of concrete are adding to the heat-island effect that can make life difficult in Southwestern cities for plants and people alike. A study in Tucson found that a blue paloverde (*Parkinsonia florida*) could reduce soil temperature on cement and gravel up to 68°F! This creates a much more favorable environment for plant growth in the direct vicinity of a tree. On a larger scale, a study in Phoenix showed that increasing tree canopy cover from 0 to 25 percent in a neighborhood could reduce the total average temperature by 7.9°F, making our cities much more inviting for all organisms.

On the opposite temperature extreme, trees help prevent radiant heat loss from the soil in wintertime, which in

A microclimate created by a south-facing wall and well-placed rocks allows a gardener to grow cacti and succulents in a high-elevation garden.

turn helps insulate other plants from nighttime lows and may be the difference between a plant receiving frost damage or not. So, one great way to mitigate the effects of heat and cold on your plants is to create a layered garden with a canopy that helps insulate plants and soils from temperature extremes.

In addition to layered plantings, it is important to consider the microclimates in your yard. Microclimates can exist on larger or smaller scales. For instance, a house located near the northern base of a mountain range and next to a wash will be in a cold-air drainage where frosts will tend to be more frequent and of longer duration. On a smaller scale, the north side of a house will tend to drain cold air and take longer to warm up because it will receive less sun than other parts of the yard. The reverse is also true. South- and west-facing exposures generally get more sun and will experience greater radiant heat. Though these types of microclimates can present challenges, they also offer opportunities to cultivate plants that would otherwise struggle in your landscape. Taking advantage of microclimates can allow you to grow species that might naturally occur just above or below the elevation of your landscape, broadening the possibilities for plant selection in your garden.

A series of basins and swales turns the challenge of gardening on a narrow hillside into an advantage as the water from a street and parking lot are diverted toward native plants that slow the flow of water and prevent erosion.

Water

It should come as no surprise that water is the most important consideration when designing a landscape in the Southwest. The relative scarcity of this important resource is what has driven the evolution of many of our native plants, and concern about water usage is one of the main drivers for gardeners to turn to native species in the first place.

Everyone wants to know how much to water their plants, but providing precise guidance is difficult because there are so many variables, from where you are gardening to what your soil is like, what plants you are growing, and what the sun exposure is in your planting area. What is essential, regardless of these factors, is that you alter your watering regimen to account for seasonal changes. A fully leafed-out tree in 100-degree heat will require much more water than a dormant one during the coldest part of winter. Flexibility is a key prerequisite of efficient irrigation. But the very best way to get a sense of your garden's watering needs is to spend time with your plants! Observing how long it takes for signs of drought stress (like wilt or leaf drop) to appear between waterings and sticking your fingers into the soil to feel how long it stays moist are both excellent ways to get a sense of the particular needs of the plants in your landscape. However, a few broad guidelines and best practices will help you get your plants off to a good start.

Stretchberry (*Forestiera pubescens*) is an appropriately sized plant for this raised garden bed.

The first thing to consider is what type of plant you are growing. Different types of plants have different root systems and water storage capabilities, which means that there is not one right way to water every plant in a landscape.

TREES AND LARGE SHRUBS

Trees and large shrubs will tend to have large root systems commensurate with the size of the plant. Their taproots reach deep into the ground to pull up moisture banked in the soil, while a widespread system of feeder roots closer to the surface extends beyond the canopy to soak up moisture as it falls. When irrigating trees and large shrubs, the goal should be to water deeply and slowly to encourage a robust root system. Shallow watering near the trunk of the plant will stunt roots and keep them from growing in search of water, meaning your plant will be more needy and susceptible to blowing over in strong winds. Waterings should be long in duration, with moisture penetrating at least 3 feet into the soil. The length of watering required to meet that criteria will depend on your soil type. For newly planted trees, water frequently, but over the course of your tree's life, gradually reduce the frequency and increase the duration to encourage robust roots that will help your plant become more drought resilient. Even for large shrubs and trees, most of the roots are near the soil surface, so as the canopy of your tree or shrub expands, you will water farther and farther away from the trunk. Watering just beyond the edge of the canopy will encourage feeder roots to keep moving outward in search of moisture. Your watering goal should be to wean your large trees and shrubs off supplemental water over time.

PERENNIALS, VINES, AND GROUNDCOVERS

Variations in root size and type can be significant between different desert plants, but typically, they will consist of one or more taproots that go straight into the soil with smaller feeder roots extending outward closer to the soil surface. Because of the smaller size of these plants relative to trees and larger shrubs, they prefer shorter-duration waterings. But your goal is still to drive the roots downward and outward in search of moisture. With these species, the relatively small canopy size means you will probably not have to substantially alter where irrigation emitters are placed, but when laying the hose or installing emitters, be sure to place them at least a few inches from the base of the plant to encourage the roots to develop fully. When planting shallow-rooted perennials, you will need to water much more frequently to get them established. You should aim to wet the soil down to 1 or 2 feet while moving your plants away from reliance on irrigation over time.

GRASSES

Grasses lack significant taproots and typically have a fibrous root system that is ideal for binding together soil and preventing erosion. The size of the root system will be related to the size of the grass but often extends at least a couple of feet into the soil, with feeder roots running parallel to the surface of the ground. Our native grasses can be divided into cool- and hot-season grasses, and this is important to keep in mind when watering. Grasses that are dormant until the onset of midsummer humidity will not need much water during winter and spring, whereas cool-season grasses will want irrigation as winter comes to a close to take advantage of their spring growth period. When watering grasses, try to wet the soil down to 1 to 2 feet.

CACTI AND SUCCULENTS

When watering cacti and succulents, it is important to remember that they store moisture in their cells, an important adaptation that allows them to bank moisture internally. The trade-off is that they have shallow, wide root systems, and even large cacti like saguaros (*Carnegiea gigantea*) have only very short taproots. The upshot of this is that long, deep waterings like you would give to a tree will be wasteful when applied to cacti that have their roots so close to the soil surface. Some succulents like desert spoon (*Dasylirion wheeleri*), beargrasses (*Nolina* spp.), yuccas (*Yucca* spp.), and ocotillo (*Fouquieria splendens*) have less moisture-storage capacity and can benefit from a greater frequency of watering than cacti or agaves. Cacti and succulents are more prone to rot than other plants, so it is especially important that water is reduced during the cold season and that these plants are in relatively well-draining soils. When watering cacti and succulents, wetting the soil down to about a foot in depth should be sufficient to reach most of their roots.

Monsoon clouds bring the promise of rain and respite from the heat of a Southwestern summer.

CHOOSING AN IRRIGATION METHOD

Before breaking ground on your new native plant garden, your first consideration should be how you will get water to your plants. There is no single right or wrong way to water, and each option comes with advantages and disadvantages. Before any plants go in the ground, you want to decide if you will be watering by hand or using a drip irrigation system. This will be important because any hose bibs or irrigation lines should be installed before planting so that water will be immediately available to newly planted specimens, and you won't have to disturb recently planted sections of your garden trenching waterlines.

Watering by hand is certainly the simplest method for irrigating, and it has the advantage of getting you out in your garden regularly. This helps you learn when your plants need water so that you can tailor your watering regimen to them as they grow and as seasons change. Watering by hose also comes with a low initial price tag and doesn't necessitate the maintenance of an irrigation system. On the other hand, hose watering is relatively inefficient in terms of time and water. Hoses can use up to 20 gallons of water a minute, and when water is puddling around a plant, a high proportion of it may evaporate before reaching the roots, especially in summertime when plants need moisture most. Take into account that as you bring more plants into your garden, hand watering can become an onerous, time-consuming task. If you plan on leaving town for any length of time during summer, missed waterings can result in the death of plants and a loss of the time, effort, and money that went into them, as well as the wildlife value they would have provided.

One compromise is to use a soaker hose. These are hoses that are capped on the end and drip water out of their whole length. Soaker hoses are ideal for circling around trees and other plants that require deep watering. Using a soaker hose is much more efficient than spraying water onto plants and will encourage deep watering that promotes a healthy root system without the expense and effort of a drip irrigation system. It is worth noting that soaker hoses often have a short lifespan in the Southwest, where thirsty critters, salty water, and punishing sun break them down quickly.

Hose watering is one of the most common ways to irrigate plants, but it can become a time-consuming task as your garden grows.

The other option is to invest in a drip irrigation system. Drip irrigation is undeniably more efficient than hose watering, in large part because water is released slowly and evenly from emitters, going directly to the roots of your plants. Drip irrigation systems can also be set up so that water is only going to the plants you choose, rather than being spread out over the entire garden, where it might go toward encouraging the disturbance-loving, non-native plants that are fond of most yards. The ability to preset irrigation duration and frequency also means that your personal time investment will be much lower with a drip irrigation system than with hand watering.

With that said, drip irrigation systems are not perfect. To begin with, the initial investment can be higher, especially if you are hiring a professional landscaper to install your system. Additionally, these systems require routine maintenance, such as moving emitters as plants grow or adding new lines as more plants are installed. These systems also inevitably break down over time and require patching or replacement to fix leaks. This is particularly true because most places in the Southwest will have hard water with a high amount of calcium that breaks down and clogs irrigation systems.

Because of the different watering needs for different types of plants, you will want to install separate lines for succulents, shrubs, and trees. This adds a level of complexity but will be helpful in the long run because you can adjust watering as needed and accommodate the different types of plants in your yard. There are many different ways to water off an irrigation line: bubblers that pour out water and are mostly used for trees; perforated tubing with holes evenly spaced along the length, which can be used to water many plants in a bed at once; or feeder lines tipped with emitters that release a preset amount of moisture measured in gallons-per-hour. Once an irrigation system is installed, any combination of these watering technologies can be integrated.

Overall, drip systems are better for water efficiency (an important factor in the Southwest) and are less complicated than they might first appear. That being said, if you don't have experience with irrigation systems, it is best to hire someone to design and install them, though you can easily gain the skills necessary for basic maintenance.

It is ultimately up to you to decide how to water your plants, but it is a factor that needs to be considered before the first plant ever goes into the ground.

Garden Design

A good landscape design gives you a roadmap to guide your efforts that may help you save the time, money, and heartache that can come with piecemeal gardening without a long-term vision. A garden design can be as simple as a drawing on a napkin created by a home gardener or as

Artistic rockwork, a meandering path, and healthy plants turn this suburban front yard into a living showcase of the ecological and aesthetic benefits of gardening with native species.

Native succulents are planted on a south-facing terraced slope, where they benefit from sun and reflected heat while their shallow root systems help bind the soil.

complex as a 3D design generated by a professional landscape architect. If you are developing a design yourself, the best way to prepare is to visit nearby botanical gardens, look for opportunities to participate in community garden tours, and take trips to view the unmatched design aesthetic of Mother Nature.

Many elements go into a good design, but one of the most important is selecting the right plant for each location. Beyond that, considerations include coordinating bloom times to ensure a long flowering season, mixing complementary flower and foliage colors, and even accounting for the shadows a plant might cast on a wall. All these factors taken together will add up to a garden that is a pleasure to spend time in.

Designing for the Whole Year

Gardens go through natural cycles, and it is unrealistic to expect them to look the same all year. Humans don't always look our best in 100-degree heat; the same goes for plants. Anticipating and appreciating these changes will give you greater enjoyment in your garden and help you eliminate unnecessary work. On the other hand, a thoughtful design can help ensure that there is almost always nectar and pollen available for insects, cover for birds and mammals, and beauty for you. If you take some time to study the flowering seasons accompanying each plant description in this book, you can select plants that will give you blooms throughout much of the year. It is also important to remember that the seeds of many plant species are food sources for wildlife, so the typical task of deadheading plants right after they flower denies urban wildlife an important resource. Even when your plants are dormant, they are still providing nesting sites for insects and birds. We have become culturally accustomed to the notion of keeping our yards "clean," but this artificial standard keeps us working constantly, expending fuel and chemicals to maintain a manicured and ecologically bereft landscape. It is our task as informed gardeners to push back against this paradigm and allow our plants to go through natural cycles of bud, leaf, flower, fruit, and dormancy, as all stages of life have value, not just those in which the plant is at its showiest.

Plant Succession

Succession is the process in which landscapes go from being disturbed to having complex species assemblages, by passing through stages from fast-growing pioneer species to slow-growing but long-lived trees and shrubs. Succession will occur in our gardens just as it does in nature, and we can plan for this to ensure our gardens continue to thrive in the long term.

Many of our native trees, shrubs, and succulents take time to reach maturity. We can compensate for this fact by

These plants are just beginning to establish in their first year after planting.

A lack of forethought has forced this homeowner to modify their roof to accommodate a saguaro planted too close to the house.

including fast-growing perennials and subshrubs to add instant color and fill in blank spaces while longer-lived plants get their roots under them. An inexpensive way to do this is to purchase a wildflower seed mix filled with species native to your area and sow this around your garden as you make your initial plantings. These wildflowers will quickly germinate, and if all goes well, they will reseed and appear year after year.

You want to plan for the ultimate spread of large trees and shrubs to protect structures and minimize the need for unnecessary and unsightly pruning. A quick look around the ecosystems of the Southwest will show that plants generally have plenty of space between them, yet it is fairly common to see gardeners overplant an area in the hopes of filling it quickly, only to have to go back and remove plants that aren't thriving because of competition and lack of space. It is also typical to see large trees planted too close to structures where their limbs or roots can cause catastrophic damage to waterlines, roofs, or walls, resulting in the need for drastic pruning or removal that negates the time invested in growing the tree in the first place.

Balancing instant and delayed gratification will ensure that our yards are beautiful and functional places at every stage of their growth and will save time and money in the long run. Long-term planning can be particularly challenging for new gardeners or those unfamiliar with the plants they are putting in the ground. It takes experience to understand how the shrub or tree purchased in a 1-gallon pot will look in 10 or 15 years. On your next hike, make note of the mature size of the plants that you see, which species

Natural ecosystems consist of groups of plants that occupy different niches and exist in a state of both competition and collaboration.

occur under trees, and which species are thriving in full sun. Because plants don't always act the same in our gardens as they do in the wild, it is also worth visiting nearby botanical gardens and arboretums to see how these plants respond to cultivation. Planning for succession helps bring our gardens more in line with natural systems and helps us differentiate between normal cycles of growth and decay or other issues that may require intervention.

Guilds

No species in nature occurs in a vacuum. A gardener who seeks to create biodiversity and functionality in their yard should base their design not just around individual plants, but around the miniature ecosystems they will be part of. Diverse habitats are made of guilds of plants that show variety in form, structure, and ecological niche and together create resilient spaces that are useful for wildlife.

Guilds are assemblages of different species that can be combined to perform mutually beneficial roles in a plant community. Perhaps the most well-known guild in the Southwest is the "three sisters," a combination of corn, beans, and squash where the corn provides a trellis for the beans, the beans add nutrients to the soil, and the squash acts as a weed suppressant. This popular vegetable gardening technique applies equally well to landscaping with native plants. For instance, a bean-family tree like velvet mesquite (*Neltuma velutina*) hosts nitrogen-fixing bacteria in its roots that enrich the soil while the tree provides shade and a trellis for vines; a wild squash like buffalo gourd (*Cucurbita foetidissima*) can be planted to cover the ground and suppress unwanted weeds; and a native grass, like giant sacaton (*Sporobolus wrightii*), can add visual structure and seed for wildlife. The options for guild combinations are nearly unlimited and can be tailored to the conditions of your yard. Some examples of guilds for different bioregions and planting situations can be found at the end of this book. These examples are far from exhaustive and are meant to get you thinking about combinations you can make for different situations in your own yard.

This Santa Rita pricklypear (*Opuntia santa-rita*) receives shade and nutrients from an overhanging foothill paloverde (*Parkinsonia microphylla*).

Pruning

We've all seen it. In neighborhoods, parking lots, and public buildings, perfectly good trees are hacked and sawed into sorry skeletons and nearby shrubs are buzzed into unnatural gumdrop shapes. Far too many trees and shrubs fall victim every day to the heavy hands of uninformed arborists or overenthusiastic homeowners who end up damaging plants beyond repair. Wherever possible, it is ideal to put your plants in the right location so that they can be left to grow without intervention. That being said, pruning is often necessary, especially in smaller urban yards or streetside plantings. Where pruning is required, it should be done in a considerate, deliberate, and informed manner with an eye toward the long-term health and form of the plant in question.

A bigtooth maple (*Acer grandidentatum*) is pruned just high enough to allow for foot traffic on a path.

This is far from a comprehensive review of pruning theory and technique, but the goal is to leave readers with an overview of common mistakes and best practices that can help guide your own pruning or allow you to better judge the work of arborists and landscapers you are considering hiring.

Set Goals

When starting a pruning project, it is helpful to have a goal in mind. In other words, why are you pulling out the saw? Perhaps you are hoping to lift a tree's canopy to make it easier for people to walk underneath. Maybe you want to thin a shrub to open up a view. Or it could be that you are trying to clear deadwood and remove weight from an ungainly specimen. These are all valid reasons to prune, but they would involve slightly different approaches. Ideally you will never prune off more than 25 percent of a tree at once, so knowing what your ultimate goal is will help you make only the cuts you need while preventing over-pruning.

The Tools

The most commonly used tool for most gardeners will be hand pruners. It is worth purchasing a decent pair and keeping a sharpening tool on hand to maintain the blade. There are two common types of pruners: anvil and bypass. Bypass pruners are what you want to use for living plants, while anvil pruners are better suited to piecing out dead branches. For branches ½ inch thick or larger, you want to switch to using loppers, which are essentially larger two-handed pruners. Most loppers are built to make cuts on branches from ½ to 2 inches thick, so beyond that, you will need to pull out the handsaw. This is generally where mistakes start getting made, but if you keep your saw sharp, then the three-cut method described here will allow you to handle most pruning around your yard. For higher branches, you can invest in a pole-saw or pole-pruner. The techniques you use will be the same, though extra caution is necessary when cutting limbs high up in a tree, as they can suddenly drop and cause serious injury or property damage. At the point when pruning work requires pulling out a ladder, it may be time to consider calling a professional or, at the very least, having a second person present, as off-ground work significantly increases the risk of injury.

The Three-Cut Method

It's worth reviewing just how to go about pruning a branch with a saw. This may seem basic, but branches are often mangled beyond repair by people not adhering to the technique outlined below. The steps to pruning a branch are as follows:

- Assess the branch and the surrounding area to ensure that no one will be hurt and nothing will be damaged by you removing the branch. You want to start with dead, diseased, or damaged branches and then move on to branches that are a hazard because of their height or because they are growing over a structure.
- Follow the branch down to the "collar," where the limb meets a main trunk; this will be the site of your final cut. Prune just outside the collar so that the tree can seal off the wound and heal quickly. Cutting too far from the collar leaves a stub that will be unsightly and give rise to many suckering branches. Cutting too flush with the trunk makes a larger wound that will be difficult to heal.
- Make sure your saw is sharp and clean so it makes effective cuts and doesn't introduce pathogens from other trees. Then, make a cut on the underside of the branch approximately 6 to 9 inches away from the branch collar. You want to cut about one-third to one-half of the way through the bottom of the branch.

The three-cut method will help you prune in a way that protects the tree and encourages quick healing from cuts.

- Cut the top of the branch just beyond the undercut. The branch will break cleanly without tearing, leaving behind a stub of the branch.
- Now that most of the weight is off the branch, make a cut parallel to the trunk and just outside the branch collar.

Common Pruning Mistakes

The two issues listed here are some of the most common and easily avoidable pruning mistakes, and simply avoiding these practices will go a long way in protecting the health of your plants.

LION-TAILING

This is the most common form of tree abuse. Lion-tailing involves removing nearly all inner or lateral branches of a tree, leaving a narrow trunk with a top-heavy pom-pom of branches and foliage like the puff at the end of a lion's tail. This is typically done on trees in public spaces, parking lots, and walkways to allow for foot or car traffic. When lion-tailing is combined with the selection of fast-growing but structurally weak hybrid and nursery-cultivar trees, the result is inevitable collapse or limb-drop during a storm. It's understandable to want to increase the canopy height of a tree, but it can be done in a more responsible way that will preserve the aesthetic value of the specimen. When raising a canopy, it is important to start early in the tree's life and make it a regular part of your maintenance regimen. This will allow you to take branches when they are smaller and prune a little at a time rather than making drastic cuts of large branches all at once. You want to start by removing the upper and lower lateral branches to take the weight off the main branches before taking a couple of the lowest main branches as needed. Good canopy lifting happens over a long period of time and starts when the tree is still young. Drastic reactionary pruning is likely to leave the tree weak, unbalanced, and more likely to cause harm. Truly the best solution for high-traffic areas is to select a naturally upright tree and design for ample space where it can grow.

SHEARING

You see a nice shrub go into a landscape, maybe a common juniper (*Juniperus communis*) or hopseed bush (*Dodonaea viscosa*). The plant begins to grow beautifully, developing a natural shape and filling out the planting space. That's when the hedge trimmers appear, and in a few minutes, the shrub goes from an elegant natural shape to a drab box, ball, or gumdrop, like a parody of a plant drawn by a cubist painter. This form of pruning not only destroys the form of the plant and prevents it from flowering but also leaves roughly chopped stems that are prone to disease or sprouting unattractive new growth that keeps you

This hybrid paloverde (*Parkinsonia* × 'Desert Museum') has been severely lion-tailed and is planted in a small area surrounded by compacted parking-lot soil. It is only a matter of time before the combination of overpruning and a notoriously unstable hybrid tree result in collapse.

These desert spoons (*Dasylirion wheeleri*) have been severely sheared at the expense of their long-term aesthetic value and health.

on the shearing treadmill. This type of pruning is particularly egregious when used on succulents like desert spoon (*Dasylirion wheeleri*) and beargrass (*Nolina microcarpa*), which will be irreparably damaged. Most often, this is done because these plants were put in the wrong place to begin with. Rather than shearing, try selective pruning. Selective pruning involves the careful removal of a small number of secondary branches, or trimming branch tips back to a main stem to control the size of a plant or encourage a more open aesthetic. Selective pruning will leave many branches intact to grow naturally and flower and will preserve the aesthetic of your shrubs. There may be instances where coppicing a plant back to the ground is warranted or even beneficial, but shearing is always a poor practice.

What about Mistletoe?

Mistletoe (*Phoradendron* spp.) has quite a reputation. It is either presented as a symbol of romantic love or the scourge of all plants, gradually sucking the life from your trees. This is a narrative that needs to be corrected, because if you are gardening for biodiversity, then mistletoe may not be an enemy but an asset. Mistletoe is a plant that has evolved to get some of its sustenance by parasitizing other plants; that is indisputable. However, it is worth considering that mistletoe has been growing on host trees in the Southwest for millennia, and yet those trees are still here. This implies that in healthy ecosystems, mistletoe and the trees it parasitizes exist in an equilibrium that has sustained them both over evolutionary time. If mistletoe infestations become unmanageable in landscapes, it says less about mistletoe than it does about the unhealthy state of trees stressed by drought, urban heat, and poor maintenance.

So, what are the benefits of mistletoe? To begin with, mistletoe is the sole larval food source for the great purple hairstreak butterfly (*Atlides halesus*), a lovely insect and a fantastic pollinator. In addition, the berries are an essential food source for phainopeplas (*Phainopepla nitens*), a charismatic bird with jet-black plumage and a distinctive crest. This bird eats mistletoe berries and deposits the seeds onto tree limbs. Insects use the blooms as a nectar source, and many birds like to use the stems for nesting material. Studies have even shown that the presence of mistletoe can lead to increased diversity of plants and animals around parasitized trees. Mistletoe does not kill healthy trees, and it has even been shown to extract less water from its host during times of drought rather than more. At worst, mistletoe is a minor nuisance in healthy specimens, and infestations only become problematic in trees that are already unhealthy, where mistletoe is more of a symptom than a cause of tree mortality.

To attract wildlife, leave mistletoe on your trees and focus on keeping your trees healthy in the long term. If you do still feel the need to remove mistletoe, keep in mind that no chemicals have been shown to be effective that don't also damage the tree. You can manually remove mistletoe by cutting its brittle stems, but it is almost certain to reappear with time. To effectively remove mistletoe, you must remove whole limbs below the infection point, and if you do that every time mistletoe appears on your tree, then you may end up doing more harm than good.

Water Harvesting

This section builds off the previous discussion of watering techniques to describe how passive and active water-harvesting infrastructure can be integrated into the design of new or already established gardens. Active water harvesting means capturing and storing water using gutters and cisterns. Passive water harvesting is the installation or enhancement of contours, basins, and swales to strategically direct water that falls onto a landscape. Water-harvesting infrastructure should be considered as you design your landscape so that you can install it before or during planting or leave space for future installation of cisterns or basins.

It is important to note that water harvesting is subject to different rules and statutes in different states and municipalities of the Southwest. For instance, the state of Arizona

A water-harvesting cistern is placed onto a concrete foundation to hold its weight. Virgin's bower (*Clematis ligusticifolia*) has been planted on a trellis to screen the tank.

Passive Water Harvesting

This is the easiest way to begin water harvesting in your landscape and can be done at a variety of scales. At its most basic level, passive water harvesting involves planting trees and shrubs into small depressions that hold water and prevent runoff while placing rocks to spread flow. More intensive practices can involve digging a network of basins and swales that hold water and keeping overflow on-site with shrubs, grasses, and trees placed in and around the basins to allow their roots to soak up extra moisture. Succulents and cacti can be placed on the mounds made from the dirt taken out of these basins. At its most complex, passive water harvesting might include using heavy machinery to build lengthy swales with overflows into lower basins, curb cuts to bring in water from the street, and terraces to spread water across slopes. Passive water harvesting is not rocket science, and simple structures can be installed by almost any gardener. The key is to observe the flow of water across your landscape following rains throughout the year and install structures accordingly. For bigger installations or infrastructure on difficult landscapes such as steep slopes,

allows gardeners to collect and store as much rainwater as they can and use it indoors and outdoors for everything from watering plants to flushing toilets. The city of Tucson, Arizona, even offers a rebate of up to $2000 to encourage home and business owners to install water-harvesting infrastructure. On the other hand, the state of Colorado only allows gardeners to harvest up to 110 gallons of water in cisterns or barrels, and this water can only be used outdoors. Knowing the laws around water harvesting in your area will help you avoid issues with local government while helping you capture and use as much water as possible.

A chicane slows traffic along a residential street while directing water into a basin that's been planted with native trees and shrubs. A block like this can harvest thousands of gallons of water in a single rain event.

An outdoor shower has a French drain that delivers water to a nearby basin. A creosote bush (*Larrea tridentata*) takes advantage of the moisture and emits its pleasant fragrance when the water hits it.

it is worth consulting with a professional designer or landscaper. Classes and workshops on water harvesting are becoming more common and may be offered by your local agricultural extension office, water company, or one of the many nonprofits in the Southwest dedicated to watershed preservation and restoration.

Passive water harvesting can be both functional and aesthetically pleasing. Make your basins into flowing, organic shapes and place rock artfully into the basin edges to accentuate their forms. Use the extra dirt to create mounds for more rot-prone plants and set rocks into the soil to check erosion and slow runoff. This is a great way to take advantage of rainfall, but it's worth remembering that rainfall is highly seasonal and often largely absent in the late spring and early summer when plants become most drought-stressed. The year-to-year variability of precipitation in the Southwest means that passive water-harvesting systems may not always provide sufficient moisture for your plants, so additional irrigation may be required in times of drought.

One way to supply additional water to your yard is to use gray water. Gray water is the "wastewater" from showers, sinks, and washing machines. In general, this water is perfectly suitable for use on landscaping plants and can significantly lower your potable water inputs onto your landscape. One very simple way to harvest gray water is to handwash dishes over a plastic container that can be emptied onto plants outside. Appliances and fixtures such as washing machines and sinks can also be plumbed to run directly out to plants in your landscape via buried PVC pipes or an aboveground pool hose that can be moved around to deliver water where it is needed. Since an average washing machine may use around 20 gallons of water per load and most households wash multiple loads per week, this method can cover a large portion of your garden's water needs, helping trees and shrubs thrive without resorting to turning on the hose.

Another increasingly popular landscape feature in the Southwest is outdoor showers, which are ideal for the warm season. An outdoor shower can be plumbed directly into an adjacent garden bed, while the structure for the shower can be used as a trellis for vines that will help blend it into the landscape.

Options for passive water harvesting are nearly limitless, and when landscape features are combined with modified appliances and fixtures, much or all of your landscape's watering needs can be met without using potable water. This is an increasingly important consideration in the drought-prone Southwest.

Active Water Harvesting

Active water harvesting is the installation of gutters and cisterns to store rainwater for future use. Gutters and cisterns can come with a high initial price tag, but the amount of water captured can account for a significant amount of the water needed for a landscape. Calculating the potential rain capture of your roof involves a fairly straightforward calculation, given that roofs are by their very nature impermeable surfaces that shed the vast majority of water that falls. First, multiply the length of your roof by its width to get the total square footage. Every square foot can capture about .62 gallons of water, so a 1000-foot roof can capture 620 gallons of water per 1 inch of rain. Knowing the average rainfall in your area will allow you to make a rough estimate of what the water-harvesting potential of your roof is.

A screen prevents debris from getting into a water-harvesting cistern, where it can rot and clog pipes.

A good water-harvesting system will consist of gutters feeding into a storage container. Storage systems can range from an old 50-gallon barrel to large tanks or underground cisterns capable of storing thousands of gallons. There are a few things to keep in mind. First, you want an opaque container that prevents the sun from shining directly on the water and spurring algal growth. Additionally, it is important to have a filter that will prevent organic matter from getting into your storage container, a sealable lid for your container to keep out mosquitos, and an overflow pipe to prevent your tank from overfilling. Some people recommend the installation and use of a first flush valve that allows you to siphon off the first bit of rain that falls to avoid any pollutants from your roof getting into your stored water supply.

The next step is deciding how you will get the stored water to your plants. Many people simply use gravity and pressure to push water through a hose bib located near the bottom of the tank. This is sufficient for plants that are close to the tank, but for plants farther away or uphill, a pump may be needed to create enough pressure to deliver the water where it is needed.

The amount of water you harvest and save has to be weighed against the amount of water your landscape requires, plus the rate of evapotranspiration (ET) in your area. ET is the measure of how much water evaporates from the soil and is transpired from plant canopies over a given time. Across the Southwest, average ET rates will be several times higher than the average rainfall, so in times of drought it may be that even active water harvesting will not be sufficient to meet the irrigation needs of your landscape.

More detailed information about water harvesting can be found in the resources section at the end of this book. But suffice it to say that, with a combination of techniques in place, the use of potable water can become a last resort rather than the first option for maintaining plant health.

Mulching

One of the most beneficial things you can do for your garden is to integrate mulch. Mulch is a broad term for material laid over the soil to shade the ground, add organic matter, and suppress weeds. The most common form of mulch in most desert gardens is a thick layer of rock called crushed granite. This is waste rock from mining operations that has become popular in Southwestern landscaping because it gives a yard a "clean" aesthetic. Unfortunately, in this instance, *clean* is a euphemism for *lifeless*. Decomposed granite yards are practically ovens, with the rocks reaching temperatures many degrees hotter than the air, resulting in excessive reflected heat that can cause even the toughest plants to wither in summer. Simply doing away with the widespread crushed-rock yard will go a long way toward cooling our neighborhoods.

Allowing leaf litter to accumulate under trees and shrubs is a free and easy way to mulch your garden.

Rock mulch can be appropriate for landscapes, especially those heavily featuring cacti and succulents, but a good rock mulch will involve stones of different sizes arranged in a natural way. For instance, large stones can be set into the soil and placed as a decorative element that can also reflect some heat at frost-sensitive species, while smaller rocks can be used to prevent erosion and create moist pockets for plant roots. Well-placed rock mulch can help your garden attain a natural aesthetic while creating helpful microclimates and leaving open spaces where birds can scratch at the dirt for their meals.

Another option is to use wood chips. Dye-free wood chips are available at most hardware stores and can also be sourced from arborists who otherwise have to pay to drop them at a dump. Wood chips will help hold moisture and cool the soil. They will also break down over time, adding organic matter to the soil and encouraging bacteria and fungi that will form the basis of a food chain that encourages wildlife from insects on up. If wood chips are applied in too thick a layer, they can suppress the growth of perennial and annual wildflowers and can also deny birds access to insects in the soil. It is also true that wood chips don't contribute to a particularly natural-looking landscape, though this may only be an issue for native plant purists. A balance might be struck by concentrating wood chips just around planting holes and in basins while leaving open spaces and decorative large rocks in between.

The best mulch for your yard will be that which falls to the ground from your trees, shrubs, and grasses. Rather than raking up all your yard waste and throwing it into a trash bag, you will be better served by chopping yard waste into smaller pieces and using it as mulch around plants, burying it at the bottom of planting holes to aerate the soil, or adding it to your compost pile where it can break down and become fertilizer for your garden. Because many members of the bean family (*Fabaceae*) host nitrogen-fixing bacteria in their roots, the litter from trees such as mesquites (*Neltuma* spp.), ironwoods (*Olneya tesota*), and paloverdes (*Parkinsonia* spp.) is particularly valuable for nutrient-limited landscapes. This litter also stimulates the growth and proliferation of beneficial fungi and bacteria. Simply leaving fallen leaves, flowers, and fruits in place or raking them around planting beds can save the money and effort involved in buying bags of wood chips or spreading loads of rock, while also improving your soil and holding moisture in place.

Distant scorpionweed (*Phacelia distans*) and desert lupine (*Lupinus sparsiflorus*) are among the many spring wildflowers that appear following good winter and spring rains in the Sonoran Desert.

Annuals

This book primarily focuses on perennial plants because these are what gardeners will generally find in nurseries year-round. That being said, annuals make fantastic additions to a landscape, adding bursts of seasonal color. Because they are so short-lived, it is often not economical for nurseries to grow or customers to purchase native annual plants, so the best way to encourage annual species is through sowing seeds rather than planting starts. Ideally, seeds should come from the closest reputable source, or if gardeners are confident in their identification abilities, harvested nearby. Seed harvesting from around neighborhoods or private properties should only be done with permission. It is essential that you know and understand local plant protection laws and that you do not illegally or irresponsibly harvest seed from wild populations.

In the Southwest, there are typically two types of annuals: those responsive to winter and spring rains and those that grow with the heat and humidity of summer monsoons. Sowing seeds of both these types of annuals will give gardeners more flowers throughout the year and help them hedge their bets in case one of the rainy seasons fails to materialize, which is often the case in the unpredictable climate of the Southwest.

Sunflowers (*Helianthus annuus*) are common along roadsides following summer rains.

SPRING ANNUALS	SUMMER ANNUALS
bladderpod (*Physaria gordonii*)	desert thorn apple (*Datura discolor*)
chia (*Salvia columbariae*)	devil's claw (*Proboscidea parviflora*)
Coulter's globemallow (*Sphaeralcea coulteri*)	fleabanes (*Erigeron* spp.)
desert chicory (*Rafinesquia neomexicana*)	lemon beebalm (*Monarda citriodora*)
fleabanes (*Erigeron* spp.)	morning glories (*Ipomoea* spp.)
flor de San Juan, yellow desert evening primrose (*Oenothera primiveris*)	scarlet bugler (*Penstemon barbatus*)
jewelflower (*Streptanthus carinatus*)	southwestern cosmos (*Cosmos parviflorus*)
lesser yellowthroat gilia (*Gilia flavocincta*)	summer poppy (*Kallstroemia grandiflora*)
lupines (*Lupinus* spp.)	sunflower (*Helianthus annuus*)
Mexican gold poppy (*Eschscholzia californica* var. *mexicana*)	cinchweed (*Pectis papposa*)
purple owl clover (*Castilleja exserta*)	tahoka daisy (*Machaeranthera tanacetifolia*)
scorpionweeds (*Phacelia* spp.)	wild zinnia (*Zinnia peruviana*)

Where to Source Native Plants

An increasing number of nurseries and private sellers are growing native plant species. Learning how to select native plants at the nursery is a skill. Don't be afraid to ask where the nursery sources their seed or cuttings. Unfortunately, it is not uncommon to see disreputable growers poaching plants, such as cacti and agaves, from their habitat. Supporting nurseries with ethical and responsible collecting protocols while boycotting those with suspect practices will hopefully discourage profit-motivated plant thieves.

You want to avoid plants that are under-rooted from being sold too early or root-bound from sitting in a pot too long. Both situations tend to result in plants that struggle when put in the ground. Under-rooting will be indicated by plants that wiggle around in the container, and when you remove the plant from the pot, most of the soil will fall right off. A root-bound plant will have a low soil level from dirt washing out over time and roots that circle around the pot. An ideal plant will be steady in the pot with roots that fill the container without excessive circling.

Avoid trees and shrubs that have been over-pruned in their containers. Keep in mind that the plant you are purchasing in a container will mature to look very different than it does now, and over-pruning too early in the plant's life can prevent it from attaining a natural shape as it grows. You also want to steer clear of plants with obvious signs of disease, as you may take a pathogen home with you that can spread to other plants. Some plants, such as agaves, are particularly prone to disease in cultivation and require close inspection.

The Southwest is home to many good native plant nurseries with horticulturists available to offer guidance.

Insect infestation is often a turn-off for nursery customers, and in some instances, rightly so. However, since many native plant gardeners are buying plants specifically to feed wildlife, a lack of any insect blemishes can be a sign that pesticides have been applied to a plant. Ask about the pest-management practices at the nursery before buying plants as a pollinator food source.

The best place to purchase plants will always be a locally owned nursery. The horticulture industry in the Southwest is full of highly experienced and dedicated individuals who are happy to share their knowledge of, and passion for, native plants. Supporting these nurseries keeps money in your community, contributing to economic as well as environmental sustainability.

Planting

Once you have your new plants, it's time to get them off to a good start. Best practices for planting can vary somewhat depending on where you are in the Southwest, but a few general rules can help get you started.

The timing of planting is important to consider. Putting in new plants in late spring or early summer means they will go into the ground just as temperatures become punishingly hot. The best planting time for most plants is the late summer and early fall monsoon season, so that your plants can take advantage of the moisture and humidity that often comes this time of year while giving them several months to establish before the return of the hot season. In colder

The profiles in this book will describe the natural habitats of each plant as well as its horticultural characteristics and needs.

parts of the Southwest, or for more frost-sensitive species, planting in the early spring just after the last frost can also work well, so long as the plants are closely observed and regularly watered as you move into summer.

Here is a review of the steps you should take to plant for success.

- Regardless of soil type, it is a good practice to dig your planting hole at least two to three times the width of the root ball in the pot. Because most plant roots run parallel to the soil surface just a few inches underground, this ensures that the roots of your new plant will be able to quickly spread out and take up moisture. The depth of the hole should be the same as the root ball of the plant in the pot.
- Pour enough water into your planting hole to fill the hole, and allow this water to drain. Not only will this give you a good idea of how well your soil drains, it will also help settle the soil and create a moist and humid environment for the roots of your plant.
- Knock your plant out of the pot; don't yank on the stem. If the plant is stuck in the pot, you may have to cut it out. Set your plant carefully into the hole, making sure that the collar of the plant is flush with, or just above, the soil line. If you are planting a tree or large shrub, then a flare at the base of the trunk should be visible. If your plant is extremely root-bound, you may choose to tease some of the roots out or, in drastic cases, trim off the bottom portion of the root ball and/or shave the sides. Because this can lead to transplant shock, it is essential that you check on your plant regularly after using this technique, and don't be surprised if you see leaf drop as a result.
- Backfill the hole with native soil and a small amount of amendments as needed, ensuring that these are well blended. Gently tamp down the soil around your plant.
- Water deeply and continue checking on the plant regularly. If you are planting leafy plants, they may need water daily or every other day, whereas newly planted succulents may go several days between waterings.

A hole is being prepared to accommodate this Santa Rita acacia (*Mariosousa millefolia*).

How to Use the Plant Profiles

The plant profiles are divided by plant type into trees, shrubs, perennials, cacti and succulents, grasses, and vines. Each profile contains plant name, size, exposure, water needs, flowers, field guide, and description.

◀ Left to right from top left corner: verdin (*Auriparus flaviceps*) on graythorn (*Condaliopsis divaricata*); painted lady butterfly (*Vanessa cardui*) and Thompson's beardtongue (*Penstemon thompsoniae*); Putnam's cicada (*Platypedia putnami*); Fremont wolfberry (*Lycium fremontii*); tarantula hawk wasps (*Pepsis sp.*) visiting Gregg's acacia(*Senegalia greggii*); carpenter bee (*Xylocopa sp.*) on scarlet gilia (*Ipomopsis aggregata*); white-lined sphinx moth (*Hyles lineata*) and Hooker's evening primrose (*Oenothera elata*); devil's claw (*Proboscidea parviflora*); Sonoran coachwhip (*Masticophis flagellum*) dining on a spiny lizard (*Scelopora sp.*)

Plant name: Each profile lists the scientific name (the binomial classification of genus and species) as well as the common name in both English and Spanish where available. This book attempts to use the most up-to-date nomenclature for each plant, though taxonomists and horticulturists are not always on the same page, and two or more taxonomic names are often in use. For plants where confusion is likely, synonyms are listed. This section also features symbols indicating the wildlife drawn to each plant.

- Bats
- Bees
- Beetles
- Birds
- Butterflies
- Caterpillars
- Desert tortoises
- Flies
- Hummingbirds
- Moths
- Wasps

Size: The approximate height and width of the plant at maturity (in feet)

Exposure: The preferred sun exposure of the plant is defined as one of the following;

- full sun (will tolerate sun all day in winter and summer)
- partial shade (prefers or will tolerate some shade from a structure or filtered light from an overhanging tree for a few hours a day)
- shade (tolerates shade several hours a day whether from a structure or densely foliated tree or shrub)

Water needs: Water needs at maturity are divided into the following three categories. It is important to note that a plant that covers a broad elevational or geographic range may have "low" water needs in one part of its range and "medium" in another.

- low (can get by with very infrequent or no additional watering once established)
- medium (requires some supplemental irrigation, even at maturity, to grow and flower to its full potential, especially during the dry season, though it may need little or no water during the cool months and rainy season)
- high (will definitely require supplemental irrigation even in the cooler months and rainy season, depending on rainfall amounts)

Flowers: This is the approximate flowering season of the plant, though it is important to keep in mind that this season depends on a variety of factors, including climatic, soil, water, and exposure conditions.

Field guide: This is a brief description of the natural range of the species; it references the biotic communities listed earlier in this book. Sometimes, this section will reference states or particular cities, but the biotic communities are used frequently to help you consider the range of the plant separate from political boundaries. The elevation range is provided as well and works as a proxy for cold and heat tolerance. A plant that grows naturally at 10,000 feet will be very cold-tolerant, while one that grows at 500 feet will handle high heat.

Description: This concise summary of each plant's traits includes recommendations for planting conditions and companions, as well as a description of specific wildlife that each plant attracts.

Trees

Trees are the centerpiece of any landscape, and proper placement and maintenance will play an important role in determining the aesthetic and wildlife value of your garden in the long term. Before planting a tree, go look at mature specimens in a variety of situations from public gardens and arboretums to streetside plantings and natural settings where possible. This research will allow you to assess which species will be right for your space, and save you time and trouble in the long run. A poorly placed tree can be an expensive nuisance and an actual hazard, while a well-placed tree can keep your home cool in summer and warm in winter, enrich your soil, and act as a nurse plant to your garden, while serving as a hub for wildlife activity in your yard. A well-planted tree can be a lifelong friend that grows with you and your family, gaining aesthetic and ecological value over time. A tree is the ultimate investment in the future of your garden.

Abies concolor – Pinaceae

white fir (pino blanco)

30–50′ × 10–20′

full sun, partial shade

medium to high water needs

flowers May–June

Field guide: A common component of Montane Conifer Forest and Subalpine Conifer Forest in the Southwest, white fir is found between 5000 and 10,000 feet.

Description: Makes a stellar landscape specimen in cool, high-elevation gardens. Well-draining acidic soil is a must, and pruning is discouraged as this tree will achieve a lovely shape, with dense foliage and a rotund base, with no intervention. Light bark and a bluish cast to the flat needles make it quite distinctive, and this tree shines as the centerpiece of a garden bed surrounded by montane trees, shrubs, and flowers such as bigtooth maple (*Acer grandidentatum*), orange gooseberry (*Ribes pinetorum*), and showy fleabane (*Erigeron speciosus*). Small mammals and birds will eat the seeds, and the dense foliage makes an ideal nesting site for a variety of bird species. The wood of this species is fairly brittle, so plant well away from structures and avoid overwatering, which will cause weak new growth.

Acer grandidentatum – Sapindaceae

bigtooth maple (palo de azúcar)

20–50′ × 10–20′

full sun, partial shade

high water needs

flowers April–May

Field guide: Thrives in woodlands and along stream edges in the Southwest's mountains, where it acts as a pioneer species, growing in clearings opened by fires or blowdown of larger trees. This species is found in Riparian Corridors in Montane and Subalpine Conifer Forest between 4500 and 7000 feet. As other trees grow, bigtooth maple recedes, and it is less common in mature forests.

Description: An upright, neat tree, bigtooth maple functions well as a sidewalk planting or a focal specimen in smaller yards where larger trees would be inappropriate. Surround your maple with shrubs and wildflowers such as orange gooseberry (*Ribes pinetorum*), yarrow (*Achillea millefolium*), and Woods' rose (*Rosa woodsii*). When bigtooth maple has its foliage, it is an ideal place for nesting birds. In fall, the leaves turn a brilliant orange-red before dropping, which makes these trees especially striking either lining a pathway or as a single specimen against a backdrop of evergreens.

Acer negundo – Sapindaceae

boxelder maple (acecinte)

15–25′ × 15–25′

full sun, partial shade

medium to high water needs

flowers March–May

Field guide: Grows in almost every part of the United States but in the Southwest is restricted to streamsides and moist drainages. In our region, they are found between 3000 and 7500 feet, where Riparian Corridors cut through Great Basin Conifer Woodland and Montane Conifer Forest habitats.

Description: This fast-growing, midsized tree is identified by its foliage with three to seven leaflets and stunning orange-and-gold fall color. With its reputation for having brittle wood, boxelder maple is best planted away from structures in an informal hedge or a cluster around a basin fed by rain runoff and gray water. The papery, winged seed coats may attract neatly patterned black and red boxelder bugs, though these are less prevalent in the Southwest than they may be in other parts of the country. Interplant with water-loving species like deergrass (*Muhlenbergia rigens*) and willows (*Salix* spp.) for a miniature riparian habitat in your backyard. Despite these trees being primarily wind-pollinated, bees are attracted to the nondescript blossoms.

Bursera microphylla – Burseraceae

elephant tree (torote)

6–10′ × 5–8′ (bigger in Mexico)

full sun

low water needs

flowers June–July

Field guide: Grows in Lower Colorado River Sonoran Desertscrub and Arizona Upland Sonoran Desertscrub habitats, on sunbaked slopes between 500 and 3000 feet.

Description: An odd plant that tends to grow as a misshapen shrub or warped tree at the northern extent of its Arizona range, where frost is a limiting factor. Like its plant-family relatives frankincense and myrrh, its sap and foliage give off a tart, piney aroma that infuses the air when plants are watered or pruned. The trunk is enlarged at the base and typically gives rise to multiple sinuous, mahogany-colored branches with bark that peels off in thin layers like sunburnt skin. Elephant trees are extremely drought-tolerant, but occasional supplemental watering in the dry months of early summer will keep your plant lush and green. The tiny star-shaped blossoms attract minute bees and dainty butterflies, and pollination results in a dusky purple fruit that splits to reveal a single, red, bird-attracting seed. Grow this as the centerpiece of a cactus and succulent garden, against a west-facing wall that gets oven-hot in summer, or nestled into a terra-cotta pot that can be moved or covered in winter to protect from frost.

Chilopsis linearis – Bignoniaceae

desert willow (mimbre)

15–25′ × 15–25′

full sun

medium water needs

flowers April–September

Field guide: Found in Arizona, New Mexico, and southwestern Utah in the sandy edges of seasonal washes; Riparian Corridors in the Sonoran, Mojave, and Chihuahuan deserts; and into Semidesert Grassland and Interior Chaparral habitats, between 500 and 5500 feet.

Description: Despite its resemblance to the weeping willows popular in Southeastern gardens, desert willow isn't related to the water-hungry true willows and will thrive in hot gardens with a modicum of irrigation. This species sports bell-shaped pink, white, or purple flowers that attract pollinators like hummingbirds, butterflies, and carpenter bees and are a good food source for desert tortoises. The foliage is also a larval food source for moths like the rustic sphinx (*Manduca rustica*), and in summer, it's not uncommon to see fat green caterpillars munching on the leaves. An excellent choice for planting near basins and swales. This tree will thrive with gray water from washing machines, sinks, or outdoor showers. Two varieties are found in the Southwest: var. *arcuata*, in the western portion of the species' range, sports thin arched leaves and white or light pink blooms; var. *linearis*, with straight leaves and darker purple blossoms, is found in New Mexico and Texas. Try to purchase plants grown from seed collected in your area (instead of trademarked nursery cultivars).

Celtis reticulata – Cannabaceae

netleaf hackberry (cumaro)

20–25′ × 20–25′

full sun, partial shade

medium water needs

flowers March–May

Field guide: Found between 2000 and 6000 feet on canyon margins and in moist meadows, in habitats ranging from the deserts through Semidesert and Great Basin Grassland up to Madrean Evergreen Woodland and Great Basin Conifer Woodland.

Description: Though it prefers seasonally moist soils, hackberry is appropriate for almost any yard and will grow well with minimal irrigation. It's popular in Southwestern landscaping due to its lack of spines and the dense shade created by its canopy. In winter, it sheds its rough, egg-shaped leaves, exposing the knobby iron-gray bark that provides year-round visual interest. Most important, it attracts a wide array of birds with its cantaloupe-flavored orange berries and sturdy branches, which make splendid nesting sites. Dozens of avian species nest in or feed on netleaf hackberry, making these trees ideal for birders. The inconspicuous flowers attract bees and butterflies; several butterfly species, such as the hackberry emperor (*Asterocampa celtis*) and American snout (*Libytheana carinenta*), use the foliage as a caterpillar food source. Netleaf hackberry is a stellar tree for low spots, basin edges, or beds fed by gray-water outlets. Interplant with desert honeysuckle (*Anisacanthus thurberi*) and three-leaf sumac (*Rhus aromatica* var. *trilobata*) to create a dynamic habitat for birds and pollinators.

Eysenhardtia orthocarpa – Fabaceae

kidneywood (palo dulce)

6–15′ × 6–15′

full sun, partial shade

low water needs

flowers April–September

Field guide: A subtropical species with a wide distribution in Mexico, kidneywood barely reaches the United States in southeastern Arizona and southwestern New Mexico. Seen on open grassy ridges and woodland edges in Semidesert Grassland and Madrean Evergreen Woodland between 1500 and 5500 feet.

Description: This distinctive tree is one of the loveliest species of the sky islands that span the US–Mexico boundary. Limited in size by frost at the northern extent of their range, they can form a respectable canopy in favorable microclimates—near a heat-reflecting wall, for instance. The thornless branches sport lime-green foliage that becomes lush with summer monsoons before turning yellow and dropping with the first frost. Leaves feed the caterpillars of Arizona hairstreak (*Erora quaderna*), ceraunus blue (*Hemiargus ceraunus*), and gray hairstreak (*Strymon melinus*) butterflies. The white flowers appear in late spring and persist through summer on spike-like inflorescences that exude an alluring honey aroma and attract both humans and pollinators. Because of its modest size, kidneywood is a great candidate for smaller yards, patios, or container plantings.

Fraxinus velutina – Oleaceae

velvet ash (fresno)

20–30′ × 20–30′

full sun

medium to high water needs

flowers March–May

Field guide: Velvet ash can be found in Riparian Corridors between 3000 and 7000 feet, in habitats ranging from Semidesert and Great Basin Grassland to Madrean Evergreen Woodland and Montane Conifer Forest.

Description: One of the friendliest trees of the Southwest, velvet ash grows into an upright, thornless shade tree with a lovely round canopy that makes it ideal for public plantings. The leaves are divided into glossy leaflets, and the bark is carved into serpentine furrows. Nondescript flower clusters appear before the first leaves in spring and cast their pollen to the wind; the wildlife value of this species lies not in the flowers but in the leaves and twigs. Two-tailed swallowtail butterflies (*Papilio multicaudata*) and great ash sphinx moths (*Sphinx chersis*) eat velvet ash foliage, and the inner bark is a food source for the western Hercules beetle (*Dynastes grantii*), the largest beetle found in the Southwest. However, this plant needs consistent moisture to thrive, so regular irrigation, rain runoff, or gray water will be required for this species to live up to its potential.

Hesperocyparis arizonica and *H. glabra* – Cupressaceae

Arizona cypress
smooth Arizona cypress (sabino)

25–50′ × 10–35′

full sun, partial shade

medium to high water needs

flowers November–March

Field guide: Found between 3000 and 7500 feet along canyon margins in Madrean Evergreen Woodland, Interior Chaparral, Great Basin Conifer Woodland, and Montane Conifer Forest habitats.

Description: Experts debate whether cypresses found near the Mexican border and those found in the central highlands of the Southwest should be considered two separate species. For the home gardener, this may be too fine a distinction, as these fast-growing trees have a similar form regardless of where they occur in the region. Broad but neatly pyramidal with drooping branch tips and rounded seed cones, Arizona cypress thrives in moist, mid-elevation canyons. The overlapping scales of its foliage emit an agreeable fragrance when crushed. These large trees are ideal nesting sites for raptors, and their seeds are an essential food source for squirrels. Best planted where water is consistently available, they will do well on the edge of a basin fed by gray water from a sink or washing machine. It's crucial to place these trees where their large size and robust root system will not threaten building foundations or sidewalks. Heat and bark beetles are potential issues for these trees, especially at low elevations.

Juglans major – Juglandaceae

black walnut (nogal silvestre)

30–35′ × 30–35′

full sun, partial shade

medium to high water needs

flowers March–May

Field guide: Occurs between 3500 and 7000 feet along moist canyon edges and marshy bottomlands in Riparian Corridors.

Description: These stately trees can become massive with age, creating a dense shade canopy. Many plants fail to thrive when planted too close to walnut trees, likely due to walnuts' release of growth-suppressing allelopathic chemicals and direct competition for light and water. This phenomenon is especially prevalent in soils with low organic matter, such as very sandy areas. Hackberries (*Celtis reticulata* or *C. pallida*), however, do well under walnuts; consider planting these species in a guild, and experiment with other shrubs around your walnut to see what works. The large yellow fruits and the nuts inside are a nutritious food source for birds and small mammals—and they will stain your skin yellow, hinting at their historic use as dye. Several moths, including the splendid royal (*Citheronia splendens*) and small emperor (*Saturnia pavonia*), eat the leaves. An outstanding shade tree for upland gardens that have a gray-water outlet or a low spot that stays wet for a few days after a storm.

Juniperus deppeana – Cupressaceae

alligator juniper (huata)

20–30′ × 10–20′

full sun, partial shade

low to medium water needs

flowers February–March

Field guide: Occurs in canyons or along rolling hills and rocky slopes in upland habitats like Semidesert Grassland, Madrean Evergreen Woodland, Interior Chaparral, Great Basin Conifer Woodland, and Montane Conifer Forest, across an elevational range from 4000 to 7000 feet.

Description: Round and dense with a lovely blue-green tint to its needle-like foliage, this is a fantastic addition to upland Southwestern landscapes. The easiest way to differentiate this juniper from other species is the dark bark that splits into scales rather than strips, giving it the look of alligator skin. The blue berries (technically cones) look like tiny Christmas ornaments decorating this evergreen tree in fall. Combine alligator juniper with oaks (*Quercus* spp.), mountain mahogany (*Cercocarpus montanus*), and three-leaf sumac (*Rhus aromatica* var. *trilobata*) to form a dense garden border. The foliage gives off a heady fragrance, and some caterpillars like the doll's sphinx moth (*Sphinx dollii*) and juniper hairstreak butterfly (*Callophrys gryneus*) will use it as a food source. Alligator juniper is extremely cold-tolerant and great for mid- to high-elevation gardens but is not appropriate for most low-desert landscapes.

Juniperus monosperma – Cupressaceae

oneseed juniper (tascate)

15–20′ × 15–20′

full sun

low water needs

flowers February–April

Field guide: Our most desert-adapted juniper, this species can be found from the upper limits of the Sonoran Desert into conifer forests, spanning grassland and woodland habitats in between. Oneseed juniper is seen between 3000 and 7000 feet and overlaps in range with Utah juniper (*J. osteosperma*); these two species can be difficult to distinguish, especially due to their penchant for hybridization.

Description: A dense, many-trunked tree that is roughly pyramidal in outline. The bark sheds in long, thin, gray-brown strips that contrast with the deep green foliage, which consists of overlapping scales daubed with whitish exudate. Our most drought- and heat-tolerant native juniper, oneseed is still better suited to upland gardens than those in desert cities like Phoenix and Tucson. Pollen-bearing individuals can release vast clouds of yellow pollen in spring, and the fruit-bearing plants sport mealy, blue-gray seed cones. Makes an excellent focal planting on a well-draining mound surrounded by shrubs; also works as a background planting in a hedge, where its sweet seed cones will attract a variety of birds and small mammals.

Juniperus scopulorum – Cupressaceae

Rocky Mountain juniper (sabino)

25–30′ × 25–30′

full sun, partial shade

medium water needs

flowers April–May

Field guide: This plant's range extends from northern Mexico to southern Canada, and in the Southwest, it is a montane species found mostly north of the Mogollon Rim. Most often associated with Riparian Corridors and adjacent slopes between 5500 and 9000 feet, this species can be seen in Great Basin Conifer Woodland, montane scrub and conifer forests, and Subalpine Conifer Forest habitats.

Description: Thrives as an understory planting but will tolerate full-sun exposure in more temperate parts of the Southwest. Horticulturally, this species is most notable for its languidly drooping branch tips and the icy-blue cast to its foliage, which imparts an elegant melancholy to this tree. This relatively fast-growing juniper has a single upright trunk, and the taste of its berries (technically seed cones) is a big draw for wildlife, so there is much to recommend it in high-elevation Southwestern gardens. Interplant in a grove with pinyon pine (*Pinus edulis*) and Gambel oak (*Quercus gambelii*) for an aesthetically pleasing landscape with high wildlife value.

Lysiloma watsonii – Fabaceae

feather tree (tepeguaje)

15–20′ × 15–20′

full sun

low water needs

flowers April–June

Field guide: In the United States, feather tree is known from only one mountain range in southern Arizona; it becomes more common in Sonora, Mexico. Look for this species between 2500 and 4500 feet in the Rincon Mountains east of Tucson, where Arizona Upland Sonoran Desertscrub and Semidesert Grassland habitats meet.

Description: A subtropical specimen with fragrant blossoms, lacy foliage, and no thorns, this is an ideal species for convincing skeptical neighbors of the beauty of native plants. Frost sensitivity means this tree is only suitable for low-elevation gardens where its dense canopy, low water needs, off-white puffball flowers, and long mahogany-colored seedpods are desirable. Feather trees are typically multi-trunked and spreading, with semi-evergreen foliage that provides good wildlife cover, top-notch shade, and food for large orange sulphur (*Phoebis agarithe*) butterflies. Try planting feather tree on the south- or west-facing portions of a yard, where it will offer cooling shade in summer but allow light to penetrate in winter when its foliage has thinned. The microclimate created by feather tree is ideal for underplanting with shade-loving plants like violet wild petunia (*Ruellia nudiflora*) and leadwort (*Plumbago zeylanica*).

Neltuma glandulosa (syn. *Prosopis glandulosa*) – Fabaceae

honey mesquite (mezquite)

15–20′ × 15–20′

full sun

low to medium water needs

flowers April–May

Field guide: Found from the Pacific coast to the Gulf of Mexico in the Mojave, Sonoran, and Chihuahuan deserts and Semidesert and Great Basin Grassland, from just above sea level to around 5000 feet.

Description: Ranges broadly across the Southwest, overlapping with velvet mesquite (*Prosopis velutina*), though one or the other species will typically be dominant in a given area. Often a dwarfed shrub in open situations, it becomes a respectable tree with ample water. Honey mesquite's drooping branches, which are especially pronounced in large specimens, resemble a weeping willow. The trunks are often twisting and curved rather than neat and upright, and the broadly spreading branches can become heavy and prone to breaking in high winds. The foliage is a food source for Palmer's metalmark butterflies (*Apodemia palmerii*) and mesquite clearwing moths (*Carmenta prosopis*). Another native mesquite in the Southwest, the screwbean mesquite (*Strombocarpa pubescens*), tends to be a smaller tree with an upright trunk and beans that look like spiraling screws. Mesquites are notoriously thorny, so wear good gloves and a thick long-sleeve shirt when pruning.

Neltuma velutina (syn. *Prosopis velutina*) – Fabaceae

velvet mesquite (mezquite amargo)

20–25′ × 20–25′

full sun

low water needs

flowers April–June

Field guide: Look for mesquite between 1000 and 6000 feet in washes and meadows or rolling slopes and mesas in Arizona Upland Sonoran Desertscrub, Lower Colorado River Sonoran Desertscrub, and Semidesert Grassland habitats.

Description: One of the most ubiquitous trees in central and southern Arizona, this species's growth habit ranges from thorny shrubs to massive trees depending on water availability. In irrigated landscapes, they often form impressive shade canopies that make ideal nesting sites for birds. Thought to be messy because they shed leaves, flowers, and beans, this tree's "litter" has a high nitrogen content and is an excellent no-cost mulch that supports the growth of other shrubs, trees, and perennials. The beans have a tasty brown-sugar flavor, making them a favorite food for human and non-human desert dwellers. Hybrid and South American mesquites have been introduced into the nursery trade but often have less structural integrity and longevity than our native plants; avoid plants of uncertain provenance. Take advantage of the temperature-regulated microclimates created by mesquites to plant sun- or heat-sensitive shrubs that will benefit from the overhanging canopy.

Olneya tesota – Fabaceae

ironwood (palo fierro)

20–30′ × 20–30′

full sun

low water needs

flowers April–June

Field guide: The natural range of this species closely follows the boundaries of the Sonoran Desert, where it can be found just a couple hundred feet above sea level near Yuma, Arizona, up to nearly 3000 feet in the mountains west of Tucson, Arizona.

Description: An iconic desert tree with steel-gray bark, slightly recurved thorns, and a sprawling canopy of dense semi-evergreen foliage. Aside from saguaros (*Carnegiea gigantea*), ironwoods are generally the tallest plant found in the Sonoran Desert. Their canopies create islands of fertility where their nutrient-rich leaf litter and dappled shade encourage the germination and growth of a host of other plants. Place in well-draining locations, as dense clay soils will further slow the already leisurely growth rate of this species. Lovely pink-and-white flowers provide a striking contrast to the vibrant yellow of foothill paloverde (*Parkinsonia microphylla*), which bloom around the same time, and these two trees can be included in a landscape together to create an eye-catching floral display in spring.

Parkinsonia florida – Fabaceae

blue paloverde (paloverde azul)

20–25′ × 20′

full sun

low to medium water needs

flowers March–April

Field guide: Largely restricted to dry washes and sandy drainages in Lower Colorado River and Arizona Upland Sonoran Desertscrub habitats into the Mojave Desert. Can be found from just above sea level to nearly 4000 feet.

Description: This showy specimen with striking blue bark is found across southern Arizona, where it is the official state tree. The natural form can be quite bushy, with branches hanging all the way to the ground, but thoughtful pruning can create a lovely shade specimen that makes an ideal nurse plant for young cacti and succulents. The brilliant golden flowers are showier than those of foothill paloverde (*Parkinsonia microphylla*) and appear earlier in spring; planting both species in a landscape will stretch out the bloom season and ensure a greater supply of food for pollinators. Because its natural habitat is washes and drainages, placing blue paloverde near a basin and providing supplemental water during the hot season will ensure faster and more robust growth; use a light touch when pruning to avoid creating an unbalanced shape that will be susceptible to storm damage.

Parkinsonia microphylla – Fabaceae

foothill paloverde (paloverde)

15′ × 15′

full sun

low water needs

flowers April–May

Field guide: A Sonoran Desertscrub species found in both the Lower Colorado and Arizona Upland subdivisions, this species is best suited to rocky hillsides and wash edges between 500 and 4000 feet.

Description: One of the toughest trees found in the deserts of the American Southwest. Its small, drought-deciduous leaves and photosynthetic lime-green bark are signs of how well-adapted this species is to the challenging growing conditions found in southern Arizona. Acts as an important nurse plant for saguaros (*Carnegiea gigantea*) and other succulents. The flowers are a striking yellow that brightens up natural and urban landscapes from April through May. The foliage feeds paloverde webworm (*Faculta inaequalis*) caterpillars, the blossoms attract a range of pollinators, and the beans are a reliable food source for all desert dwellers, including humans and other mammals. Use foothill paloverde as the centerpiece of a cactus garden with fishhook barrel (*Ferocactus wislizeni*) and Santa Rita pricklypear (*Opuntia santa-rita*).

Picea pungens – Pinaceae

blue spruce (abeto)

30–60′ × 20–30′

full sun, partial shade

medium water needs

flowers May–July

Field guide: Found between 6000 and 10,000 feet in Montane Conifer and Subalpine Conifer Forest. Look for blue spruce in high ranges like the Sangre de Cristo, San Juan, and White mountains. This species is absent from the southern parts of Arizona and New Mexico, where it is replaced by Engelmann spruce (*Picea engelmannii*).

Description: A slow-growing tree with stiff, sharp needles and a classic Christmas-tree form that makes it a delightful landscape element for cool, high-elevation Southwestern gardens. Prefers acidic soil but is not picky about drainage and will tolerate a variety of exposures and the shade of other trees, making it a reliable understory planting. The bluish foliage contrasts nicely with darker trees and shrubs like hairy mountain mahogany (*Cercocarpus breviflorus*) and creambush (*Holodiscus discolor*). Currants (*Ribes* spp.) are a great addition under blue spruce, where the little berries add a pop of color. Blue spruce is often visited by Cooley's spruce gall adelgid (*Adelges cooleyi*), a tiny insect that chews on the branch tips and forms galls that resemble seed cones except for their needle-like appendages.

Pinus edulis – Pinaceae

pinyon pine (piñon)

25–30′ × 20–25′

full sun

medium water needs

flowers March–April

Field guide: This quintessential species of the upland and mountain Southwest is found on rocky slopes, mesas, and dry flats between 5000 and 7000 feet. Look for it in Great Basin Conifer Woodland, Great Basin Montane Scrub, and Montane Conifer Forest habitats.

Description: This relatively small tree has needles in pairs, emits an intoxicating resinous fragrance, and produces cones with seeds favored by birds and mammals, making it an excellent choice for high-elevation wildlife gardens. Aside from feeding scrub jays (*Aphelocoma* spp.) and squirrels (*Sciuridae* spp.), the seeds have long been an important food source for people in the Southwest, and this tree is worth including in any upland ethnobotany or native food garden. In drier parts of Arizona, *Pinus edulis* is replaced by the more drought-tolerant single-needle pinyon (*Pinus* x*kohae*) and, near the international boundary, by border pinyon (*Pinus cembroides*). All these trees share the traits of very slow growth; a dense, rounded crown; and short, stout needles. These trees have suffered from the extreme drought and accompanying bark beetle infestations of the last few decades; keep watch for signs of insect damage.

Pinus strobiformis – Pinaceae

southwestern white pine

30–50′ × 20′

full sun, partial shade

medium water needs

flowers June–July

Field guide: Found on high slopes in Montane Conifer Forest and close to timberline in Subalpine Conifer Forest habitats, at elevations ranging from 8000 to 10,000 feet. The northern portion of its range overlaps with that of limber pine (*Pinus flexilis*).

Description: Grows happily on saddles, peaks, and rocky slopes in Arizona, New Mexico, and Colorado. The natural form is pyramidal with five-needle bundles on flexible stems. This coniferous tree has become quite popular in landscaping, as evidenced by the numerous cultivars found in nurseries, which range from weeping groundcovers to neat balls. None of these cultivars can top the wild plant with its sturdy, elegant shape. This species is appropriate for high-elevation landscapes in the northern portions of the area covered by this book. Plant in well-draining soil and provide ample water during the growing season. Slow-growing, but older specimens are impressively stately.

Pinus ponderosa – Pinaceae

ponderosa pine (pino ponderosa)

40–50′+ × 25–30′

full sun

low to medium water needs

flowers May–June

Field guide: Primarily found in Montane Conifer Forest, where it prefers dry slopes between 5000 and 9000 feet.

Description: The iconic tree of the American West, ponderosa pine covers millions of acres. Two subspecies are found in the Southwest: the Rocky Mountain ponderosa (var. *scopulorum*) in the northern portion of the region and Arizona pine (var. *arizonica*) in the borderlands of Arizona, New Mexico, and Sonora; select individuals from seed stock that is appropriate to your area. Grows into an upright spear with plated bark and lateral branches that form a conical canopy. Mature specimens can reach well over 100 feet, though typical cultivated plants will be closer to between 30 and 50 feet. In montane landscapes, it can be planted in groves or used to anchor a garden of upland shrubs and wildflowers. Watch for bark beetle damage that takes the form of little piles of reddish dust and small bore holes in the bark. These beetles can introduce fungi that may cause the decline and eventual death of the tree.

Platanus wrightii – Platanaceae

Arizona sycamore (aliso)

30–60′ × 30–60′

full sun

high water needs

flowers April–May

Field guide: Restricted to Riparian Corridors between 2000 and 6000 feet where seasonal or perennial flows can support them.

Description: The Arizona sycamore is a rebuke to the idea that there's no fall color in the Southwest. When temperatures begin to cool in the canyons of Arizona and New Mexico, the hand-shaped leaves of sycamore trees turn a rich, rusty red that contrasts with their paper-white bark. Gardeners should be aware that a sycamore tree requires a significant water commitment, meaning they should be used sparingly and will not be appropriate for all landscapes. With ample irrigation, sycamores form large, beautiful trees that make perfect nesting habitat for raptors and other large birds. The blooms take the form of fuzzy spheres drooping down from the foliage, and the seeds come enclosed in papery, winged seed coats, but they will be difficult to see on mature trees towering above your garden.

Populus fremontii – Salicaceae

Fremont's cottonwood (álamo)

60–90′ × 60–90′

full sun

high water needs

flowers March–June

Field guide: Grows along Riparian Corridors and lake edges from near sea level at the Colorado River to around 7000 feet in the mountains near Sedona, Arizona. In New Mexico and Colorado, this species is replaced by the very similar eastern cottonwood (*Populus deltoides*).

Description: A large tree that grows to dizzying heights. The leaves quake in summer breezes, making a lovely and calming percussive sound emblematic of hot summer afternoons. When temperatures drop, the leaves turn bright yellow before falling. The trees remain bare until spring. The high branches of these trees are a preferred nesting place for large birds of prey and the foliage provides food for numerous caterpillars, including viceroy (*Limenitis archippus*) and mourning cloak (*Nymphalis antiopa*) butterflies and cottonwood dagger (*Acronicta lepusculina*) moths. This tree is only appropriate for those rare Southwest gardens with abundant water or a natural stream, as it requires constant moisture to thrive.

Populus tremuloides – Salicaceae

quaking aspen (álamo temblón)

20–50′ × 10′

full sun, partial shade

high water needs

flowers March–June

Field guide: Thriving in woodland clearings and mountain meadows, this is an early successional species that appears following disturbances such as fires or logging. Found in the Southwest between 6000 and 10,000 feet in Montane Conifer and Subalpine Conifer Forest habitats.

Description: These remarkable organisms form clonal colonies comprising genetically identical trees with a common root system. These clusters can become massive with age, with some colonies estimated to be tens of thousands of years old. As they grow, their white bark becomes mottled and covered with little dots that resemble wooden eyeballs watching passersby. The yellow-green leaves tremble in the breeze, giving off a sound like gently running water and casting intricate, dancing shadows. Only appropriate for planting in higher-elevation gardens where its robust growth makes it a suitable windbreak or screen, ideally far enough away from structures that the spreading root system will not create an issue. Hosts an astonishing number of butterfly and moth caterpillars, including those of viceroy (*Limenitis archippus*) and western tiger swallowtail (*Papilio rutulus*) butterflies and great ash sphinx moths (*Sphinx chersis*). The spreading root systems and resultant sprouts of aspens can take over an area; regular removal of new stems may be part of your routine maintenance.

Psorothamnus spinosus – Fabaceae

smoke tree (palo de humo)

15–25′ × 10–15′

full sun

low water needs

flowers June–July, October–November

Field guide: Grows in sandy washes between 200 and 2400 feet in Lower Colorado River Sonoran Desertscrub.

Description: In the scorching desert washes of western Arizona, wispy gray sprays of branches rise like puffs of smoke from sandy soil. These are the spiny limbs of one of the Southwest's most extraordinary arboreal oddities, the smoke tree. This is no shade tree; its flexuous, ashy branches are generally leafless, becoming impenetrably dense with age. Few trees are better for hot, exposed succulent gardens with soils that don't hold moisture. In early summer, smoke tree teems with flowers, robing the plant in gorgeous indigo blossoms. The fruit is a small beaked legume coated with orange glands that give off a scent like crushed citrus peel. Place on a well-draining mound surrounded by other low-desert dwellers like rush milkweed (*Asclepias subulata*), California barrel cactus (*Ferocactus acanthodes*), and desert marigold (*Baileya multiradiata*).

Quercus arizonica – Fagaceae

Arizona white oak (encino blanco)

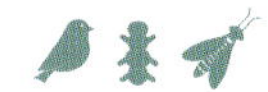

25–35′ × 25–35′

full sun, partial shade

medium to high water needs

flowers March–May

Field guide: Primarily associated with Riparian Corridors, Arizona white oak is found in Semidesert Grassland, Madrean Evergreen Woodland, Interior Chaparral, and Montane Conifer Forest habitats between 4000 and 8000 feet.

Description: Identification of this species relative to other oaks can be difficult, especially given the notoriously promiscuous nature of oaks and their tendency to hybridize. Look for leathery leaves, ranging from sharply toothed to perfectly oblong, and light-colored bark divided into furrowed strips. This powerhouse wildlife tree encourages avian and mammal visitors and provides food for charismatic oak tussock moths (*Orgyia* spp.). White oak requires access to water to thrive, so it is best planted near a basin or swale where it is easy to irrigate and will receive runoff during rainstorms. These trees are evergreen and provide year-round shade, and though slow-growing, they are extremely long-lived, with individuals often topping several hundred years.

Quercus emoryi – Fagaceae

Emory oak (encino negro)

20–30′ × 20–30′

full sun

low to medium water needs

flowers April–May

Field guide: Most prominent in Interior Chaparral, Semi-desert Grassland, and Madrean Evergreen Woodland communities between 3500 and 7000 feet in central and southern Arizona and New Mexico.

Description: This lush oak casts a deep shade that cools the ground and creates favorable conditions for shade-tolerant grasses like pinyon ricegrass (*Piptochaetium fimbriatum*) and vine mesquite grass (*Hopia obtusa*), which can form dense carpets. Emory oak prefers slightly acidic soils, so it will tend to look chlorotic (yellow leaves) in alkaline desert gardens but is otherwise fairly heat-tolerant and evergreen in all but the worst drought years. Humans and wildlife prize the acorns of this species for their relatively low tannin content, and they have been a staple food source in the Southwest for millennia. The dark green, glossy foliage and widespread canopy make this one of the most stately of all Southwestern trees.

Quercus hypoleucoides – Fagaceae

silverleaf oak (encino colorado)

15–20′ × 10–20′

full sun, partial shade

medium water needs

flowers April–June

Field guide: Favors canyon margins, gravelly slopes, and rock outcroppings in Madrean Evergreen Woodland and Montane Conifer Forest habitats between 4000 and 9000 feet.

Description: Silverleaf oak is arguably the most elegant of all Southwestern oaks. This species has two-toned leaves that are moss green on top and frosty white below. It is typically an upright specimen with a single prominent trunk that branches a few feet above the ground and develops into a rounded crown. In a garden, it is best suited to well-draining soil on the outer edge of a basin or swale, in a mixed grove of other tree and shrub species. Silverleaf oak is extremely cold-tolerant, and despite its restricted natural range in the Southwest, it will perform well in mid- to high-elevation gardens throughout the region.

Quercus gambelii – Fagaceae

Gambel oak

20–25′ × 15′

full sun, partial shade

medium water needs

flowers April–June

Field guide: Occupies rocky slopes and clearing edges in Great Basin Conifer Woodland, Montane Conifer Forest, and Subalpine Conifer Forest habitats from 6000 to 10,000 feet.

Description: This widespread oak is distinctive with its rounded, wavy leaf margins and shrubby habit when young. It will resprout if damaged, forming thickets of flexible branches that may require thinning. However, it can grow into a respectable midsized tree where fire and chainsaws are absent. This species performs well as an understory tree in a shady backyard and fits well into a planting of grasses, wildflowers, and small shrubs. It also attracts a cornucopia of insects, from moth caterpillars to gall-forming wasps and sawflies to the larvae of the Colorado hairstreak butterfly (*Hypaurotis crysalus*). Some of these insects leave galls on the leaves, but these are to be encouraged because of their visual interest and the added diversity of insects they bring to a yard. The flowers are inconspicuous and wind-pollinated, but the acorns are a lovely mahogany color.

Salix gooddingii – Salicaceae

Gooddingʼs willow (sauce)

15–30′ × 10–15′

full sun, partial shade

high water needs

flowers March–June

Field guide: Always associated with Riparian Corridors but occurs across a broad elevational range, from near sea level in the Sonoran Desert along the Colorado River to over 8000 feet in Montane Conifer Forest.

Description: This lovely riparian tree evokes mountain creeks and valley washes swollen with monsoon rains. Narrow leaves droop from branches with deeply furrowed bark, and the catkins (spike-like flowers) develop out of the leaf axils in spring. These trees can vary substantially in habit: some individuals are dense shrubs, and others form massive umbrella canopies. Willows like to have wet feet, so they will not be appropriate for every landscape in the Southwest; however, for stream stabilization and gray-water outlets, Goodding's willow can be an excellent choice. It is a host plant for various butterfly and moth species, such as viceroys (*Limenitis archippus*) and Io moths (*Automeris io*), and the blooms attract native bees.

Sapindus saponaria var. *drummondii* – Sapindaceae

western soapberry (jaboncillo)

15–20′ × 10′ (forms colonies)

full sun, partial shade

medium water needs

flowers May–August

Field guide: Found from 2500 to 5500 feet in meadows, drainages, and stream banks from the Sonoran and Chihuahuan deserts into Interior Chaparral, Madrean Evergreen Woodland, and Great Basin Conifer Woodland habitats.

Description: This small tree does well in basins fed by rain runoff or gray water. Can be confused with the similar-looking black walnut but is smaller and often spreads by runners that sprout from the root system to form thickets. Cream-colored flowers appear in summer on pyramidal spikes at the branch tips. Pollen- and fruit-bearing flowers may be on the same plant but are often on separate individuals. Numerous butterfly, bee, and beetle species visit the blossoms, and the leaves are a food source for soapberry hairstreak (*Phaeostrymon alcestis*) caterpillars. The pistillate flowers become translucent orange fruits that glow when lit by the setting sun. These saponin-rich fruits are inedible to humans but when soaked in water, they create a soapy lather that inspired the plant's common name. Despite the saponin, some birds, such as cedar waxwings (*Bombycilla cedrorum*), eat the fruit. In fall, the leaves turn orange and yellow before dropping; the fruits remain hanging like ornaments on the scaly gray bark.

Senegalia greggii – Fabaceae

catclaw acacia (uña de gato)

10–15′ × 10–15′

full sun

low water needs

flowers April–October

Field guide: Most often encountered in the Sonoran and Mojave deserts, but also occurs in and around sandy washes in Chihuahuan Desertscrub, Semidesert Grassland, Great Basin Grassland, and Interior Chaparral habitats from nearly sea level to around 5000 feet.

Description: People from densely forested northern climates often view desert trees with distaste or outright contempt for their small size and prickly nature. Perhaps no other desert tree suffers more disdain than the catclaw acacia, neglected in favor of more charismatic species or derided for its vicious, claw-like thorns. But these perspectives underestimate this tree. Its dense canopy forms a favorite roosting spot for Gambel's quail (*Callipepla gambelii*), and the pale bottlebrush blossoms entice bees and butterflies. Best of all, catclaw acacia is a bean-family tree with nitrogen-fixing bacteria in its roots, so the leaf, flower, and seed litter enrich soil if left in place as mulch. Mature specimens can become admirable shade trees, but the many low, spiny branches suggest planting along a fence or property line in a hedge where human traffic will be minimal. Thick gloves and a shirt you don't mind ripping are essential for planting and pruning this tree.

Shrubs

Once your trees are in place, you can begin planting shrubs in your landscape. These are the plants that will help define the boundaries of your garden and determine what types of wildlife you will attract. Shrubs are perennial plants with woody stems that tend to stay under 15 feet and rarely form a canopy. Outside of that definition, the plants presented here take a diversity of forms, ranging from low, spreading mounds to upright hedges and everything in between. The large number of shrubs in the Southwest means that this section only scratches the surface of possibilities available to gardeners. When selecting shrubs, think about function; in other words, what purpose will this plant serve? Is it a hedge to screen a view? A showy centerpiece in a bed of wildflowers? Will you use them as a foundation planting against a structure or scatter them through the yard to add height and texture? Are you planting for the flowers or to attract the caterpillars of your favorite butterfly? There are myriad ways to use shrubs in your landscape, so take the time to think about the outcomes you want for your yard. Be sure to make a realistic assessment of how much space you have available, so that you can place your shrubs in appropriate locations where they will not need excessive pruning later on.

Abutilon palmeri – Malvaceae

Palmer's mallow

4′+ × 3–4′

full sun, partial shade

low water needs

flowers March–November

Field guide: Found in the rugged mountains of the Sonoran Desert around Phoenix, Arizona, and Organ Pipe Cactus National Monument, where it thrives on gravelly slopes and rocky drainages between 1000 and 3000 feet.

Description: Palmer's mallow is a robust, fast-growing shrub with an upright habit and a long bloom season. The fuzzy leaves are a larval food source for a wide array of butterflies and moths, including the rufous-banded crambid moth (*Mimoschinia rufofascialis*), the Arizona powdered skipper (*Systasea zampa*) and common streaky skipper (*Celotes nessus*) butterflies, and others. Pollinators delight in the apricot-colored, cup-shaped blossoms that are a particular favorite of longhorn (*Melissodes* spp.) and chimney (*Diadasia* spp.) bees, who roll around drunkenly in the flowers, covered in pollen. Following pollination, the blossoms give rise to cheese-wheel-shaped seedpods that split to reveal multiple chambers as they dry. The seeds inside are highly viable and will contribute a steady supply of volunteers to your landscape. The large, soft leaves and long bloom season of Palmer's mallow make it a solid choice to break up the monotony of a sheet-metal fence or block wall.

Acaciella angustissima – Fabaceae

prairie acacia

3–5′ × spreads

full sun, partial shade

medium water needs

flowers April–September

Field guide: Favors rocky slopes in Semidesert Grassland and Madrean Evergreen Woodland habitats from 2500 to 6500 feet.

Description: Prairie acacia spreads by its root system to form dense colonies of upright branches with ferny foliage and puffball blossoms. The aggressive root system of this species can make it competitive in gardens, so give it room to spread and use this trait to your benefit in areas where erosion prevention is a goal. The lacy leaflets of prairie acacia are a larval food source for species like the mesquite stinger moth (*Norape tener*) along with mimosa yellow (*Pyrisitia nise*), Mexican yellow (*Eurema mexicana*), and acacia skipper (*Cogia hippalus*) butterflies. Grasses like tanglehead (*Heteropogon contortus*) and sideoats grama (*Bouteloua curtipendula*) are ideal companions for this shrubby acacia in cultivated landscapes.

Aloysia gratissima – Verbenaceae

bee brush (jazminillo)

6–8′ × 4–5′

full sun, partial shade

low to medium water needs

flowers April–September

Field guide: A subtropical plant that reaches the northern limit of its range in southern Arizona and central Texas, where it is found on rocky slopes in Semidesert Grassland habitats from 3000 to 4000 feet.

Description: An upright shrub with evergreen foliage, this plant functions well as a hedge or a focal point in a pollinator garden. The spikes of small white blossoms exude a tantalizing honey fragrance and are as attractive to bees and butterflies as they are to humans. The seeds are relished by a variety of small backyard birds like lesser goldfinches (*Spinus psaltria*) and verdins (*Auriparus flaviceps*), so it is best not to cut it back after flowering. This plant is quite drought-tolerant and will handle frosts down to at least 18°F. In habitat, bee brush is frequently found in association with velvet mesquite (*Neltuma velutina*), and this dynamic duo can be replicated in your landscape.

Ambrosia dumosa – Asteraceae

white bursage (chicurilla)

2–3′ × 3–4′

full sun

low water needs

flowers year-round

Field guide: Grows in some of the harshest portions of the Sonoran and Mojave deserts, where few other perennial shrubs can survive. Look for this plant from just above sea level to 3000 feet on gravelly slopes and sandy wash edges.

Description: Bursages are often passed over as landscape plants, both because of their rugged appearance and the windblown pollen of their flowers, which can aggravate allergies. But these plants are an integral element of Southwest desert landscapes, where they bind soil and act as nurse plants for seedling saguaros (*Carnegiea gigantea*). White bursage is a larval host for the olive-shaded bird-dropping moth (*Ponometia candefacta*) and Smith's geometer moth (*Animomyia smithii*). For a foolproof drought-tolerant planting, combine white bursage with creosote (*Larrea tridentata*) and teddy bear cholla (*Cylindropuntia bigelovii*), and seed the soil with annual wildflowers that will respond to rains and add seasonal pops of color.

Anisacanthus thurberi – Acanthaceae

desert honeysuckle (cola de gallo)

4–6′ × 3–4′

full sun, partial shade

low water needs

flowers March–June, October–November

Field guide: Found along sandy washes from 2000 to 5000 feet in Arizona Upland Sonoran Desertscrub, Chihuahuan Desertscrub, and into Semidesert Grassland.

Description: This plant is easily distinguished by its masses of orange, red, or yellow blooms, with hummingbirds busily darting back and forth between them. Its upright form and lime-green foliage look great combined with Parry's penstemon (*Penstemon parryi*) and fairy duster (*Calliandra eriophylla*) against a backdrop of netleaf hackberry (*Celtis reticulata*). The leaves are a food source for caterpillars of various checkerspot butterflies, and the flowers are particularly well adapted to hummingbird pollination, so this species will be a worthy addition to any pollinator garden. Track down individuals with different bloom colors and mix them in your garden.

Arctostaphylos pungens – Ericaceae

pointleaf manzanita (manzanilla del monte)

4–6′ × 6–8′

full sun, partial shade

medium water needs

flowers February–June

Field guide: This is the most broadly distributed of the Southwest's manzanitas; its range covers the northwest to southeast trending mountain ranges that run from Utah and Nevada to southern Arizona and New Mexico. Found from 3000 to 8500 feet in Semidesert Grassland, Interior Chaparral, Madrean Evergreen Woodland, and Great Basin Conifer Woodland habitats, up to Montane Conifer Forest.

Description: Often found in clearings created by fire, as its seeds generally require fire or smoke to germinate. This makes pointleaf manzanita tricky to germinate and grow and not particularly common in the nursery trade. Cultivation is also complicated by the fact that manzanitas in habitat ally with a particular type of fungal partner and prefer acidic soils, neither of which is typically found in Southwestern gardens. These challenges are compensated for by the gorgeous red bark, delicate urn-shaped blossoms, and plump rust-colored fruits that make this a distinctive and highly desirable landscape plant for mid- to high-elevation gardens.

Artemisia filifolia – Asteraceae

sand sagebrush

4′ × 4–6′

full sun

low water needs

flowers August–November

Field guide: Thrives in sandy soils and dunes in Great Basin and Chihuahuan Desertscrub, as well as Great Basin Grassland and Great Basin Conifer Woodland from 3500 to 6000 feet.

Description: The windblown, silver-blue foliage of sand sagebrush invokes images of the wide-open expanses of the Great Basin. Like other Artemisias, the thin leaves emit a pleasing minty fragrance that carries on the air following monsoon rains (or a good hose-watering). As the name suggests, sand sagebrush prefers well-draining soil and will happily grow on terraces, slopes, or raised mounds surrounded by wildflowers and succulents like narrowleaf yucca (*Yucca angustissima*) and plains pricklypear (*Opuntia polyacantha*). The flowers are nondescript and wind-pollinated, but the seeds attract quail (*Odontophoridae* spp.) and the dense growth habit makes this a good wildlife cover plant. Cold-tolerant, drought-hardy, and strikingly lovely, this is an ideal species for mid-elevation Chihuahuan or Great Basin gardens.

Artemisia ludoviciana – Asteraceae

western mugwort (chamizo cenizo)

3′ × 3–5′

full sun, partial shade, shade

medium to high water needs

flowers August–November

Field guide: A common element of sandy washes and floodplains across almost all Southwestern habitats from desertscrub to Subalpine Conifer Forest between 2000 and 9000 feet.

Description: This species is widespread across the Southwest and much of North America; it sports silver foliage on thin branches that emerge from the root system to form dense patches. Generally found in washes and drainages, where it benefits from extra moisture while binding the soil with its robust root system. After scouring floods, this is typically one of the first species to reappear. The whole plant has a strong minty fragrance and, like many other members of the genus *Artemisia*, has a history of medicinal and culinary use. Because it is primarily wind-pollinated, the flowers are very discreet. However, the seeds are eaten by small birds and the foliage is a larval food source for caterpillars of American lady and painted lady butterflies (*Vanessa virginiensis*, *V. cardui*) and a variety of moths, including owlet (*Noctuidae* spp.) and geometer (*Geometridae* spp.) species.

Artemisia tridentata – Asteraceae

big sagebrush (chamizo blanco)

4–6′ × 4–6′

full sun

low water needs

flowers June–October

Field guide: Common on open flatlands and mesas in Great Basin Desertscrub, Great Basin Grassland, and Great Basin Conifer Woodland from 4000 to 7000 feet.

Description: This is the iconic species of the Great Basin Desert, and like the saguaro (*Carnegiea gigantea*) in the Sonoran Desert or the Joshua tree (*Yucca brevifolia*) in the Mojave, it is visually and ecologically central to the landscapes it inhabits. This plant dominates broad expanses of open plains, lending a gray-blue hue to the viewshed and casting its alluring fragrance into the air following rains. Can be especially prevalent where overgrazing has removed grasses and small shrubs and where fire suppression has favored this burn-sensitive species. For a naturalistic look, plant in full sun in well-drained soil, interspersed with grama grasses (*Bouteloua curtipendula*, *B. gracilis*), yuccas (*Yucca elata*, *Y. brevifolia*), flattop buckwheat (*Eriogonum fasciculatum*), and buffalo gourd (*Cucurbita foetidissima*). In a more formal planting, it could be set in a row against an adobe-colored wall.

Asclepias linaria – Apocynaceae

pineleaf milkweed (hierba del cuervo)

3′ × 3′

full sun, partial shade

low water needs

flowers year-round

Field guide: This uncommon plant is found from 3000 to 6000 feet on gravelly slopes and rocky outcrops in Semidesert Grassland and Madrean Evergreen Woodland habitats.

Description: Of the thirty-plus species of milkweeds found in the Southwest, this is arguably the most attractive and well-suited to horticulture. Found in southeastern Arizona and extreme southwestern New Mexico, it is fairly cold-tolerant and handles the high heat of desert gardens as well. The densely packed rows of needle-like foliage are evergreen unless the plant becomes stressed. Blooms can appear at almost any time of year, forming clusters of small white flowers that attract a variety of insects such as beetles, bees, and butterflies, including monarchs (*Danaus plexippus*) and queens (*Danaus gilippus*). As a food source for monarch caterpillars, it's surpassed by other milkweeds; use species like horsetail (*Asclepias subverticillata*) or butterfly milkweed (*Asclepias tuberosa*) to attract caterpillars, and use pineleaf as a reliable nectar source. This milkweed is, however, the sole larval food source for the crimson-bodied lichen moth (*Lerina incarnata*). Requires well-draining soil and will turn yellow and drop leaves, or outright die, in heavy clay.

Asclepias subulata – Apocynaceae

rush milkweed (jumete) 

4′ × 4′

full sun

low water needs

flowers year-round

Field guide: Grows in the Sonoran Desert on sandy flats and rocky slopes between 1000 and 3000 feet.

Description: At first glance, this seems an unlikely landscaping plant; its branches resemble a gray-blue cluster of bare sticks, with loose bunches of cream-colored blooms. These blossoms attract a fascinating but misunderstood pollinator, the tarantula hawk wasp (*Pepsis formosa*). These large wasps have iridescent black and blue bodies with fiery orange wings. Despite the fearsome reputation of their sting, cases of humans being stung by tarantula hawks are quite rare. They prefer to incapacitate unfortunate tarantulas, which serve as living buffets for these wasps' developing larvae. As they feed on nectar, the wasps pick up conglomerations of pollen called *pollinia* from the undersides of the flowers and carry them to the next flower on their hooked feet. Plant this species in a cactus and succulent garden to take advantage of its remarkable drought tolerance and enjoy the sight of the beautiful tarantula hawk wasps and red milkweed beetles (*Tetraopes tetrophthalmus*) it attracts.

Atriplex canescens – Amaranthaceae

fourwing saltbush (chamizo) 

4–6′ × 5–6′

full sun

low water needs

flowers April–October

Field guide: Range encompasses almost the entire southwest at elevations of 100 to 6500 feet. This plant is most prominent on alkaline flats, open hillsides, and dunes in desert and grassland bioregions.

Description: This common shrub of the arid West thrives on alkaline, nutrient-poor soils and forms dense thickets of gnarled, woody stems and blue-gray foliage. Pollen- and seed-bearing flowers are often borne on separate plants, but this species confounds gender binaries by being able to switch from pollen to seed-bearing and vice versa based on environmental conditions. The flowers are wind-pollinated and nondescript, but the seeds are a favored food source for quail (*Odontophoridae* spp.) and other birds, who also shelter in the plant's mass of twisted stems. Salt stored in the plant cells makes the leaves taste like a desert potato chip. This foliage is a larval food source for the Atriplex case-bearer moth (*Coleophora atriplicivora*) and saltbush sootywing (*Hesperopsis alpheus*) and Mojave sootywing (*Hesperopsis libya*) butterflies. Because of its large size and evergreen habit, fourwing saltbush lends itself well to informal hedges with other low-water desert shrubs.

Baccharis salicifolia – Asteraceae

seep willow (batamote) 

6–10′ × 4–6′

full sun, partial shade

high water needs

flowers March–October

Field guide: This denizen of Riparian Corridors and small springs from 100 to 6500 feet grows from the low desert along the Colorado River to tree-lined canyons in Madrean Evergreen Woodland.

Description: Seep willow forms upright thickets of graceful stems with lance-shaped leaves. Not a true willow, this species is actually a member of the sunflower family, and in summer it breaks out into a profusion of white compound flowers that provide a veritable nectar-and-pollen buffet for pollinators of every type, who swarm busily around the blossoms. The relatively high water requirement of this species means it is best used in gardens with gray-water outlets from sinks or washing machines and is ideal as a living screen for an outdoor shower. The foliage is a food source for moth caterpillars as well as fatal metalmark (*Calephelis nemesis*) and Elada checkerspot (*Texola elada*) butterflies.

Bouvardia ternifolia – Rubiaceae

firecracker bush (cigarritos) 

3–4′ × 3′

full sun, partial shade

medium water needs

flowers March–November

Field guide: Thrives in Riparian Corridors in southern Arizona and New Mexico in Semidesert Grassland, Madrean Evergreen Woodland, and Montane Conifer Forest habitats from 3000 to 7000 feet.

Description: This elegant shrub has evergreen, glossy foliage that pops against an adobe or block wall and trumpet-shaped, lipstick-red blooms that are irresistible to passing hummingbirds. Because of its preference for canyons and streams, firecracker bush will be grateful for supplemental water and partial shade, particularly at lower elevations. Plant this species in a basin or gray-water outlet with velvet ash (*Fraxinus velutina*), deergrass (*Muhlenbergia rigens*), and desert honeysuckle (*Anisacanthus thurberi*) to create a miniature riparian area in your backyard. Damage to the leaves of this species is likely the result of falcon sphinx moth (*Xylophanes falco*) caterpillars.

Calliandra eriophylla – Fabaceae

fairy duster (huajillo)

3–4′ × 3–4′

full sun

low water needs

flowers February–April, occasionally September–October

Field guide: Found in the Sonoran and Chihuahuan deserts from 2000 to 5000 feet, on sandy wash edges and rocky slopes.

Description: This hardy shrub with showy blooms is a workhorse plant for Sonoran Desert gardens. It sports globe-shaped heads of numerous flowers with white and pink stamens that advertise their pollen to passing bees and butterflies. These blooms are a prominent part of the spring floral display in the Sonoran Desert and may give a second flush following monsoon rains. Drought and heat are no issue for this species, though the tiny green leaflets borne on thin gray branches will drop if no additional water is provided during the hot early-summer months. Because of its small size and thornless branches, fairy duster can be used in street or path-side plantings, and it also looks great set into a bed of ocotillo (*Fouquieria splendens*), fishhook barrel cactus (*Ferocactus wislizeni*), and foothill paloverde (*Parkinsonia microphylla*), where its airy texture offsets the spiny edges of those species.

Canotia holacantha – Celastraceae

crucifixion thorn (corona de Cristo)

9–15′ × 6–9′

full sun

low water needs

flowers March–September

Field guide: Found on rocky slopes and sandy wash margins at the upper edges of the Sonoran and Mojave deserts into Interior Chaparral, Semidesert Grassland, and Great Basin Conifer Woodland habitats between 2000 and 5000 feet.

Description: Of all the gnarled, spiny shrubs in the Southwest, this is one of the most impressive and intimidating. Where the low desert gives way to scattered clumps of juniper or thickets of chaparral, crucifixion thorn becomes common, casting its mostly leafless stems into the air in a twisted mass. The small cream-colored flowers occur in clusters that entice butterflies, bees, and several species of wasps, a fact that may be viewed as a negative or a positive. Pollinated blossoms give rise to sharply beaked seedpods that persist for many months. This plant benefits from good drainage and can be grown as the centerpiece of a cactus-and-succulent garden or mixed into a chaparral hedge with scrub oak (*Quercus turbinella*) and three-leaf sumac (*Rhus trilobata*). Perhaps not everyone's cup of tea, but its drought tolerance, pollinator value, and distinctive form make it a useful plant for Southwest gardeners unafraid to embrace the desert aesthetic.

Capsicum annuum var. *glabriusculum* – Solanaceae

bird pepper (chiltepin)

3–5′ × 3–4′

partial shade

medium water needs

flowers May–September

Field guide: Found between 3500 and 4500 feet in the sky islands of southeastern Arizona in Semidesert Grassland and Madrean Evergreen Woodland.

Description: Chile peppers are quintessential in Southwest summer gardens, but few people know that around 95 percent of all cultivated pepper varieties are derived from one wild species that reaches the northernmost limit of its natural range in southern Arizona. This wild pepper is known as a chiltepin, and its small spherical chilis are indispensable to the cuisine of the US–Mexico borderlands, where their fiery heat imparts an unforgettable flavor and sensation to any dish. In warmer areas, these plants can be perennials that persist for many years situated under mesquites or in other shady spots. In cooler areas, chiltepins can be grown as annuals in a vegetable garden. The leaves feed caterpillars of the tobacco hornworm moth (*Manduca sexta*), and the berries are a favorite of birds. The plant itself is highly decorative, with five-petaled white- or lavender-tinged flowers and tiny fruits that turn from green to scarlet red. This plant is the "mother of all chilies," and its inclusion in your garden will not only benefit wildlife and spice up your table fare, but will also aid in conserving a plant that is central to the agricultural and culinary past and future of the Southwest.

Cercocarpus montanus – Rosaceae

mountain mahogany (palo duro)

6–12′ × 4–8′

full sun, partial shade

medium water needs

flowers April–September

Field guide: An integral understory element of Madrean Evergreen Woodland, Great Basin Conifer Woodland, and Montane Conifer Forest communities in the Southwest from 4500 to 8500 feet.

Description: This dense shrub serves as a source of food and shelter for wildlife. The flowers are thin tubes that terminate in saucer-shaped openings with recurved white petals that are primarily wind-pollinated but still attract bees and butterflies. One of the benefits of this species is that the roots host nitrogen-fixing bacteria that improve the nutrient content of surrounding soil, making it a useful addition to nutrient-poor landscapes. Mountain mahogany seeds are topped by feathery threads that drill into the soil when they become wet, helping the seed successfully germinate. Use this species as a background planting with lower shrubs and wildflowers in front, or place it between evergreen trees like Arizona cypress (*Hesperocyparis arizonica*), alligator juniper (*Juniperus deppeana*), or pinyon pine (*Pinus edulis*).

Chamaebatiaria millefolium – Rosaceae

fernbush

4–8′ × 4–8′

full sun, partial shade

medium to high water needs

flowers July–November

Field guide: Native to the mountains of northwestern Arizona and southern Utah and Nevada, fernbush grows on rocky slopes, gravel flats, and sandy wash edges in Great Basin Conifer Woodland and Montane Conifer Forest from 4500 to 8000 feet.

Description: With feathery dissected foliage and masses of white blossoms, this is a distinctive and delightful shrub for high-elevation gardens. It is quite cold-tolerant and will handle poor soils and full-sun exposure with aplomb, acting as a striking centerpiece to a pollinator garden or a conspicuous element in a pollinator hedge. The summer-blooming flowers are borne on conical spikes that will entice butterflies and bees, while the dense semi-evergreen foliage provides effective cover for birds and small mammals. The dark green leaves of fernbush contrast nicely with the silver foliage of sand sagebrush (*Artemisia filifolia*) and banana yucca (*Yucca baccata*).

Condalia warnockii – Rhamnaceae

Warnock's snakewood (crucillo)

6–9′ × 6–9′

full sun

low water needs

flowers May–September

Field guide: Look for this species along sandy washes or on rolling hills in Arizona Upland Sonoran Desertscrub, Chihuahuan Desertscrub, and Semidesert Grassland from 1500 to 5000 feet.

Description: This underutilized plant for low-elevation gardens is typically overlooked in favor of better-known species like desert hackberry (*Celtis pallida*) and wolfberry (*Lycium* spp.), which have similar spiny forms with edible berries. You will be surprised by the massive diversity of pollinators that visit the small yellow-green flowers and the range of birds that arrive to devour and spread the purple berries. Best suited to well-draining sandy soil, this plant will grow several feet tall and wide, forming a thick hedge of acute spines that make it best to plant along a wall, well away from paths. The whorls of small leaves are a food source for Mexican silk moths (*Agapema anona*), and the impenetrable mass of branches makes excellent cover for small backyard birds like lesser goldfinches (*Spinus psaltria*). An ideal choice for wildlife gardeners who aren't afraid to embrace (so to speak) the spiny flora of the desert Southwest.

Condea emoryi (syn. *Hyptis emoryi*) – Lamiaceae

desert lavender (salvia del desierto)

6–8′ × 3–4′

full sun

low water needs

flowers year-round

Field guide: Found between 500 and 3500 feet in and around sandy washes in Arizona Upland Sonoran Desertscrub, Lower Colorado River Desertscrub, and Mojave Desertscrub.

Description: In a land of diverse botanical aromas, desert lavender stands out for its intoxicating and alluring fragrance. This member of the mint family consists of tall, spindly stems sporting pairs of small gray leaves and branches tipped by clusters of purple blossoms. Don't hesitate to place desert lavender in the hottest, most exposed parts of your landscape, but be sure to plant it in well-draining soils where it will put on height more quickly. It is also wise to site this species near walkways where chance brushes against it will elicit the calming scent of the leaves—and where the native bees and butterflies visiting the flowers will be more visible.

Dalea pulchra – Fabaceae

bush dalea

3′ × 3′

full sun, partial shade

low water needs

flowers February–May

Field guide: Look for this species in Arizona Upland Sonoran Desertscrub, Semidesert Grassland, and Madrean Evergreen Woodland habitats between 2500 and 5000 feet in the sky islands of southern Arizona and northern Sonora, Mexico. This plant is fond of gravelly slopes and rock outcrops.

Description: Bush dalea is an upright silver-foliaged shrub with small purple and white flowers that occur in globe-like clusters at the branch tips as temperatures rise from late winter into spring. The tiny leaflets are evergreen (ever-silver in this case) and contribute year-round visual interest. Among others, this stout, charming shrub attracts caterpillars of gray hairstreak (*Strymon melinus*) and Reakirt's blue (*Echinargus isola*) butterflies, which utilize the citrus-scented leaves as a food source. Dalea is one of the largest genera of flowering plants in the Southwest, so there is a whole world of lovely species to draw on for gardens at different elevations and exposures.

Dasiphora fruticosa (syn. *Potentilla fruticosa*) – Rosaceae

shrubby cinquefoil

2–3′ × 3–4′

full sun, partial shade

medium to high water needs

flowers May–October

Field guide: This plant appears north of the Mogollon Rim, where it can be found along canyon edges and woodland meadows. Look for it in Great Basin Conifer Woodland, Montane Conifer Forest, and Subalpine Conifer Forest communities from 6000 to 11,000 feet.

Description: Sure to delight high-elevation gardeners in either mass plantings or scattered in with other flowering shrubs, this plant's dark evergreen foliage and yellow flowers provide year-round visual interest and make it perfect for lining a path. Try this species in a bed edged with bigtooth maple (*Acer grandidentatum*) or plant it against a large rock that the branches can drape over, and surround it with wildflowers such as purple poppy-mallow (*Callirhoe involucrata*), long-flowered four o'clock (*Mirabilis longiflora*), and yarrow (*Achillea millefolium*). The yellow flowers occur en masse from early summer through fall, depending on elevation and aspect, and they are a preferred food source for native bee species. Because of this plant's affinity for moist habitats, be sure to provide supplemental water during the dry season.

Dodonaea viscosa – Sapindaceae

hopseed bush (tarachico)

6–8′ × 4–6′

full sun

low to medium water needs

flowers year-round

Field guide: This species is distributed globally in subtropical areas, but in the Southwest, it is only found in Arizona, where it grows on rocky slopes and canyon margins in Arizona Upland Sonoran Desertscrub, Interior Chaparral, and Semidesert Grassland habitats between 2000 and 5000 feet.

Description: This fast-growing evergreen shrub is perfect for creating hedges or screens around the margins of a yard or against a wall. The variety common to Arizona has thinner leaves than many nursery cultivars and is much more drought-tolerant and cold-resistant. Though the flowers are nondescript, the papery, winged seedpods found on fruit-bearing individuals take on a reddish blush as they mature that contrasts beautifully with the glossy green foliage. Plant with Arizona rosewood (*Vauquelinia californica*) and desert hackberry (*Celtis pallida*) to create a visually and physically impenetrable hedge. The foliage is a larval food source for the Cincta silk moth (*Rothschildia cincta*).

Encelia farinosa – Asteraceae

brittlebush (incienso)

2–3′ × 3′

full sun, partial shade

low water needs

flowers November–May

Field guide: Grows on sandy wash margins and rocky slopes from nearly sea level to 3500 feet in the Sonoran and Mojave deserts.

Description: One of the most spectacular sights in the desert Southwest is the stunning display of spring flowers that appear in years with good winter rains. These displays can be fleeting and elusive, requiring just-right conditions, but some species will reliably flower every year. In years with few blooms, brittlebush ensures that at least some flowers will be available for passing pollinators and wildflower watchers. The vibrant yellow blossoms are carried on thin stems, and the gray-green leaves are covered with a coating of fine white hairs. When fully mature and in flower, this shrub takes on a dome shape that contrasts nicely with the upright forms of chain-fruit cholla (*Cylindropuntia fulgida*) and ocotillo (*Fouquieria splendens*). The seeds of brittlebush are eaten by a variety of backyard birds that will help spread them.

Ephedra aspera – Ephedraceae

rough jointfir (canutillo)

4′ × 4′

full sun

low water needs

flowers January–April

Field guide: Found on well-draining slopes and gravelly flats from an elevation of 1000 feet in the Sonoran, Mojave, and Chihuahuan deserts into Semidesert Grassland and Interior Chaparral up to 4000 feet.

Description: Jointfirs are gymnosperms—cone-producing plants like pines and junipers—that predate the evolution of flowering plants. Despite its ancient origins, this plant is fantastically adapted to the arid desert Southwest, and its mass of reed-like stems can be seen across a broad swath of the region. This mostly leafless shrub has pollen- and seed-bearing cones on separate individuals, and despite lacking petals, the cones are showy en masse, lending the bare stems a golden cast and occasionally wafting out little clouds of gold as their pollen blows in the wind. This drought-tolerant shrub can be grown alongside cacti and succulents that share its low water requirements and tolerance of blazing sun and reflected heat. Though slow-growing and sometimes difficult to find in nurseries, it is well worth including in your garden for its distinctive form and its status in geological history.

Ericameria laricifolia – Asteraceae

turpentine bush

3′ × 3′

full sun

low water needs

flowers August–November

Field guide: Found in the Sonoran, Mojave, and Chihuahuan deserts as well as Semidesert Grassland and Madrean Evergreen Woodland habitats. You can find this common shrub on gravelly slopes, canyon walls, and rock outcrops from 3000 to 6000 feet.

Description: This plant blends into the background through most of the year, a dense mass of resinous needle-like leaves that doesn't draw much attention to itself. But come fall, when the grasses have dried tan and the wildflowers have spent their last blooms, it comes alive with an abundance of cheery yellow flowers that breathe new life into tired landscapes and attract any bees and butterflies still hanging around your yard. Its evergreen foliage and stout habit make it work well as a foundation planting, an element of a path-side hedge, or a reliable addition to a year-round pollinator garden. Another related species, rubber rabbitbrush (*Ericameria nauseosa*), is a widespread member of upland shrub communities and features similar blossoms, but on a plant with a bluish cast and longer, upright stems.

Eriogonum fasciculatum – Polygonaceae

flattop buckwheat (maderista)

3′ × 3–4′

full sun, partial shade

low water needs

flowers March–November

Field guide: Found in central and western Arizona, up to southern Utah and Nevada, and west into California, where it grows on rocky, exposed hillsides and wash edges in the Sonoran and Mojave deserts from 1000 feet into Interior Chaparral and Semidesert Grassland habitats up to 4500 feet.

Description: One of the preeminent pollinator plants of the Southwest, this species' showy clusters of white and pink flowers attract a wide variety of bees, butterflies, and even hummingbirds. The most notable visitors will be hairstreak, metalmark, and blue butterflies that delight in the pollen and nectar, while electra buckmoth (*Hemileuca electra*) caterpillars feed on the leaves. As the flowers age, they turn a rusty brown that is as lovely as the fresh blooms. This plant has been used widely in revegetation for burned or disturbed sites and can be placed in a landscape to populate rocky soils and slopes where other plants may not thrive.

Eriogonum wrightii – Polygonaceae

Wright's buckwheat

2–3′ × 2–4′

full sun, partial shade

low to medium water needs

flowers June–October

Field guide: Found across a large swath of the Southwest in habitats ranging from rocky desert slopes to wooded canyon margins. Look for it from 2500 to 7500 feet in a variety of bioregions.

Description: Less showy than its close relative, flattop buckwheat (*Eriogonum fasciculatum*), but easily matches that species in terms of its value to pollinators as a nectar and larval food source. This plant forms low, spreading masses of gray foliage that send up spindly flower stalks bearing small white blossoms tinged with a blush of rose. The flowers are favored by a variety of insects, and the long bloom season ensures sustenance for late-season pollinators as other plants go dormant for winter. Butterflies, including the lupine blue (*Icaricia lupini*), Rita dotted blue (*Euphilotes rita*), Acmon blue (*Icaricia acmon*), and Mormon metalmark (*Apodemia mormo*), use Wright's buckwheat as a larval food source, as do several species of moths. Intersperse it in a pollinator garden for the lovely contrast of its silvery foliage against green-leaved shrubs and for its role as a late-season food source for backyard insects.

Fallugia paradoxa – Rosaceae

Apache plume (póñil)

4′ × 5′

full sun, partial shade

low to medium water needs

flowers April–October

Field guide: Most common in woodland clearings or on cliff faces and rocky slopes in Interior Chaparral, Madrean Evergreen Woodland, and Great Basin Conifer Woodland and into Montane Conifer Forest from 3500 to 7500 feet.

Description: A versatile and exceedingly attractive member of the rose family, this is a sprawling plant with branches that arch gracefully toward the ground and are covered with intricately lobed leaves. The flowers are five-petaled and pearly white, but the most distinctive feature of this species is its clusters of burgundy-plumed seeds resembling a feathery headdress. Provide well-draining soil to mimic the conditions this plant occurs in naturally, and in lower-elevation gardens, consider planting on the north or east side of a structure or tree. You will certainly enjoy the parade of butterflies and bees that hang around the blossoms and the birds that visit to carry off the seeds.

Forestiera pubescens – Oleaceae

stretchberry

9–12′ × 9–12′

full sun

low to medium water needs

flowers April–June

Field guide: Prefers rocky canyon bottoms and wash edges in Interior Chaparral, Great Basin Conifer Woodland, and Montane Conifer Forest between 3500 and 7000 feet.

Description: This large shrub of the high-elevation Southwest thrives along canyon margins, forming dense thickets of glossy oval leaves and ashy gray stems. Gardeners can plant stretchberry as part of a wildlife hedge or as a specimen planting in patios or small gardens. Pollen- and fruit-bearing flowers are typically borne on separate individuals; staminate flowers lack petals and are primarily wind-pollinated, though they may be visited by bees and butterflies. The foliage is eaten by rustic sphinx moth (*Manduca rustica*) caterpillars, and the fruit-bearing flowers give rise to dark blue berries that are bitter to humans but beloved by birds. In late fall, the light green foliage turns a radiant yellow before dropping for winter. Stretchberry is deciduous, so interplant it with evergreen species like pinyon pine (*Pinus edulis*) and pointleaf manzanita (*Arctostaphylos pungens*) that can maintain visual interest through winter until stretchberry leafs out again in spring.

Frangula californica – Rhamnaceae

California buckthorn (hierba de oso)

6–12′ × 6–12′

full sun, partial shade

medium water needs

flowers April–June

Field guide: Grows as an understory shrub in Riparian Corridors and on shady slopes in Interior Chaparral and Madrean Evergreen Woodland habitats. Primarily a Californian species that makes it into Arizona and southwestern New Mexico between 3500 and 6500 feet.

Description: Because it prefers moist locales, this handsome, oval-leaved, evergreen shrub needs regular deep irrigation in the warm months to support optimal growth. The flowers are nondescript yellow-green blossoms in dense clusters that, while inconspicuous to humans, are irresistible to pollinators who swarm over them, creating an audible buzz around these shrubs in late spring and early summer. The foliage is an important food source for a wide variety of insects, including the gray hairstreak butterfly (*Strymon melinus*) and Rocky Mountain agapema moth (*Agapema homogena*). The bright red berries turn deep purple as they mature; birds readily devour them and deposit the seeds under nearby trees, explaining why this plant is so often found beneath oak canopies in habitat. An invaluable addition to mid-elevation wildlife hedges, where it will double as a visual screen.

Garrya wrightii – Garryaceae

Wright's silktassel (chichicahuile)

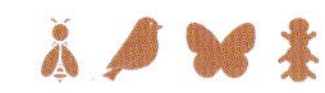

6–9′ × 6–9′

full sun, partial shade

medium water needs

flowers March–August

Field guide: A common component of Interior Chaparral, Madrean Evergreen Woodland, and Great Basin Conifer Woodland habitats, this species occurs across a broad swath of Arizona and southern New Mexico on rocky slopes from 3500 to 7000 feet.

Description: Upright, drought-tolerant, and evergreen, this plant makes for a lovely hedge, and can also be used to line driveways or paths in mid-elevation gardens. Its blue-green leaves help it stand out against other shrubs like three-leaf sumac (*Rhus aromatica* var. *trilobata*) and pointleaf manzanita (*Arctostaphylos pungens*), which can be combined with it in a garden. The foliage is a food source for the Florestan sphinx moth (*Manduca florestan*), and the fruits, borne only on plants with pistillate flowers, are a food source for birds and small mammals. The need for both pollen- and fruit-bearing individuals means that Wright's silktassel is best planted in groups, which will enhance its value as a nesting and cover plant for birds and its potential as a hedge. Benefits from a periodic deep drink during the dry months to encourage optimal growth.

Gossypium thurberi – Malvaceae

Thurber's cotton (algodoncillo)

4–10′ × 4–8′

full sun, partial shade

medium water needs

flowers July–October

Field guide: Look for this plant on wash margins and rocky slopes from 3000 to 6000 feet in Arizona Upland Sonoran Desertscrub, Semidesert Grassland, and Madrean Evergreen Woodland habitats in the sky islands of southern Arizona and northern Sonora, Mexico.

Description: One of the showiest species of the Arizona–Sonora borderlands, this plant provides pollinator value, lovely delicate blossoms, and striking fall color. It's a useful caterpillar host plant whose hand-shaped leaves provide food for painted lady (*Vanessa cardui*) and gray hairstreak (*Strymon melinus*) butterflies as well as splendid royal (*Citheronia splendens*) and owlet (*Noctuidae* spp.) moths. The summer- and fall-blooming flowers are white and cup-shaped with a pink dot near the base of each petal that helps guide the numerous insects that visit. The leaves resemble maple foliage, especially in their bright red fall color, an unusual feature for a desert shrub. The globe-shaped seedpods split to reveal fibrous strands that indicate the close relationship of this species to cultivated cotton. This resilient plant will readily reseed and contribute numerous volunteers to your garden, presenting a striking fall spectacle as the leaves change color.

Holodiscus discolor – Rosaceae

creambush

4–9′ × 4–9′

partial shade

medium water needs

flowers June–August

Field guide: Inhabits the upper elevation Montane and Subalpine Conifer Forest of the Southwest, typically growing as an understory shrub on the edges of meadows and clearings at 7500 to 10,000 feet.

Description: This species is best known for its elegant sprays of cream-colored blossoms that drape languidly from the branches, resembling the froth at the crest of a wave. The perfumed flowers entice a host of pollinators, and backyard birds devour the seeds that follow. Additionally, the lobed leaves feed brown elfin (*Callophrys augustinus*) and Weidemeyer's admiral (*Limenitis weidemeyerii*) butterflies along with blinded sphinx (*Paonias excaecata*) and spotted tussock (*Lophocampa maculata*) moths, among others. This plant's high-elevation affinity ensures cold-hardiness but limits it to more temperate gardens where it may benefit from light shade and regular deep waterings in summer. Though deciduous in winter, it's worth including in a wildlife garden or patio planting where its stunning floral display and propensity to draw in pollinators can be appreciated.

Juniperus communis – Cupressaceae

common juniper

2–6′ × 4–8′

full sun, partial shade

medium to high water needs

flowers May–June

Field guide: Found throughout much of the northern hemisphere, including the cooler mountainous portions of the Southwest in Great Basin Conifer Woodland, Montane Conifer Forest, and Subalpine Conifer Forest habitats. Look for it north of the Mogollon Rim from 8000 to 10,500 feet on rocky slopes and cliff faces.

Description: Best known as the primary flavoring agent in gin, this is an ideal landscape plant for high-elevation gardens. Unlike junipers that become sizeable trees, this species is a low-mounding shrub or groundcover perfect for lining pathways or filling in a raised planter. The stiff blue-green needles occur in clusters on the branches and host caterpillars of the juniper hairstreak butterfly (*Callophrys gryneus*). Bearing waxy berries in summer, this shrub also attracts birds, including mockingbirds (*Mimus polyglottos*) and yellow-rumped warblers (*Setophaga coronata*). Take advantage of the spreading evergreen foliage to border garden beds or knit together loose slopes. Keep in mind that there are a dizzying array of cultivars, including upright species, but that our native plants tend to be low-growing and are wider than they are tall.

Justicia californica – Acanthaceae

chuparosa (chuparrosa)

4′ × 4–5′

full sun

low water needs

flowers February–June

Field guide: Look for chuparosa on sandy wash edges and rocky foothills in the Lower Colorado River and Arizona Upland subdivisions of the Sonoran Desertscrub from nearly sea level to around 3000 feet.

Description: *Chuparosa* means "nectar sucker" in Spanish and refers to the hummingbirds that invariably visit this plant when it's in bloom. Tubular red or yellow flowers top tangled masses of blue-green stems. The blooms appear in winter and early spring when many other plants are still dormant, adding a burst of color to your landscape and ensuring a supply of nectar for resident hummingbirds. Well-adapted to the harshest conditions the Sonoran Desert has to offer, chuparosa will thrive in low-elevation gardens. It photosynthesizes when leafless, so during periods of drought, heat, and cold, the plant may shed every leaf and yet continue to thrive. When conditions improve, the leaves quickly reappear. Interplant with other drought-tolerant pollinator plants such as rush milkweed (*Asclepias subulata*) and Parry's beardtongue (*Penstemon parryi*).

Krascheninnikovia lanata – Amaranthaceae

winterfat

1-3′ × 1-3′

full sun

low to medium water needs

flowers May–October

Field guide: Covering a broad swath of the Southwest, this plant comes into contact with Joshua trees (*Yucca brevifolia*) in the sandy Mojave Desert and with ponderosa pine (*Pinus ponderosa*) in Montane Conifer Forest. It is found from 2000 to 7500 feet on sandy mesas, rolling hills, and brushy hillsides.

Description: This distinctive shrub is made up of a low mass of thin stems bearing silver-blue leaves that are considered an ideal winter forage for wild mammals and livestock. Pollen- and seed-bearing flowers are typically found on separate individuals, and though inconspicuous, the dense covering of wooly hairs that coat the inflorescences is quite decorative, especially when the hairs catch the rays of the rising or setting sun. Contrast the silvery foliage of this plant with the darker green leaves of other shrubs and grasses in a wildlife garden.

Larrea tridentata – Zygophyllaceae

creosote bush (hediondilla)

5–7′ × 5–7′

full sun

low water needs

flowers year-round (especially February–April)

Field guide: Found in the parched flats and rocky foothills of Arizona Upland Sonoran Desertscrub, Lower Colorado River Sonoran Desertscrub, Mojave Desertscrub, and Chihuahuan Desertscrub into Interior Chaparral and Semidesert Grassland habitats and ranges from close to sea level to nearly 5000 feet in elevation.

Description: One of the most drought-tolerant shrubs on the planet, this miracle of natural engineering is abundant in the desert Southwest. Creosote spreads primarily by root runners and can form large colonies, with one notable example being "King Clone," a colony in the Mojave Desert that is estimated to be nearly 12,000 years old. The oils on the leaves prevent moisture loss and exude a potent fragrance that permeates the desert following rains. More than 120 bee species use creosote for nectar, nesting, and pollen. Many other insects also visit, some of which can only survive where creosote is present, including about fifteen species of flies that form attractive, orb-shaped galls on creosote leaves, stems, and flowers. This plant is so common it is often overlooked for gardens, and yet its tolerance of extreme conditions and its status as a workhorse pollinator plant render it a necessity for any desert landscape.

Lycium berlandieri – Solanaceae

Berlandier's wolfberry (bachata)

4–6′ × 4–6′

full sun

low water needs

flowers July–November

Field guide: Found on sandy plains and rocky slopes in Arizona Upland Sonoran Desertscrub, Chihuahuan Desertscrub, and Semidesert Grassland habitats from 1000 to 4000 feet.

Description: This spiny shrub digs its substantial root system into the loose soil of rocky foothills and dry washes in search of moisture. The leaves are thinner and less succulent than those of other wolfberry species, and the bark is a distinctive dark mahogany color. This tomato family member's flowers are pollinated by insects and hummingbirds; they produce small but plump red fruits that are generally unpalatable to humans but appealing to birds and mammals. This extremely drought-tolerant species is at home in a naturalistic planting, where it will provide excellent cover for Gambel's quail (*Callipepla gambelii*) and food for Mexican silk moth (*Agapema anona*) and prominent moth (*Notodontidae* spp.) caterpillars. Give this thorny plant space to grow into its natural form, as pruning significantly reduces its aesthetic and wildlife value. Another relative, Fremont wolfberry (*Lycium fremontii*), is larger with more succulent leaves and tasty berries. Wolfberries typically lose their leaves in summer; this helps them survive until the monsoon rains.

Lycium pallidum – Solanaceae

pale wolfberry (frutilla)

3–6′ × 3–6′+

full sun, partial shade

medium water needs

flowers March–June

Field guide: By far our most widespread Lycium, this plant ranges from California to Texas and bridges several different habitat types. It's found between 3000 and 7000 feet in sandy washes and gravelly slopes.

Description: This versatile plant is suitable for a range of gardens across the Southwest. It spreads by its root system to form low, spiny thickets, a growth habit that should be kept in mind when deciding where to place it in a garden. Easily differentiated by its gray-green leaves, it also has the largest flowers and fruits of our various Southwestern wolfberries, making it a favorite edible for wildlife and humans alike. At lower elevations, pale wolfberry can be planted in the shade of overhanging trees and may benefit from being set into a basin where it can capture extra moisture. The berries are a favorite food source for birds, especially phainopeplas (*Phainopepla nitens*), who prize the fruits, and Gambel's quail (*Callipepla gambelii*), who eat the fruits and succulent leaves.

Mahonia haematocarpa – Berberidaceae

red barberry (algerita)

5–7′ × 5–7′

full sun, partial shade

medium water needs

flowers February–June

Field guide: Grows on rocky slopes and gravelly mesas in Semidesert Grassland, Interior Chaparral, Madrean Evergreen Woodland, and Great Basin Conifer Woodland habitats from 3000 to 7500 feet.

Description: While cool mornings still leave frost on windshields, red barberry explodes in a profusion of yellow flowers that garland its spiky blue-green foliage. This evergreen shrub is often encountered under velvet mesquite (*Neltuma velutina*), oak (*Quercus* spp.), and juniper (*Juniperus* spp.) or scattered in with grasses and shrubs. The dense foliage consists of spiny leaves tightly packed on the plant, making this shrub virtually impossible to walk through. Shooting out from the same nodes as the leaves are small clusters of lemon-yellow flowers on thin stems, appearing as early as February in lower elevations and brightening dormant winter landscapes. The blooms consist of a halo of yellow sepals around a central petal cup that holds the flower's reproductive parts. The flowers' many pollinators will delight insect aficionados, and the tart red berries will reward birders with sightings of curve-billed thrashers (*Toxostoma curvirostre*) and cedar waxwings (*Bombycilla cedrorum*), among others.

Mahonia repens – Berberidaceae

creeping barberry (yerba de la sangre)

6″–2′ × 4′+ (spreads by rhizomes)

partial shade, shade

medium to high water needs

flowers April–July

Field guide: Found on shady hillsides and canyon margins in Great Basin Conifer Woodland and Montane Conifer Forest habitats from 5000 to 10,000 feet.

Description: This shade-loving groundcover spreads by its root system to form large patches and is ideal for planting under trees, lining shaded walls, or filling in rock gardens. A larval food source for the barberry geometer moth (*Rheumaptera meadii*), the spiny leaves form a green mass on the ground that turns a gorgeous rust-red as winter sets in. Yellow blossoms appear throughout spring and summer and are followed by waxy-blue berries that are edible but bitter, a trait that does not deter birds who gorge themselves on the fruit. An intolerance for high heat and intense sun exposure means this species is not suitable for low-elevation gardens. However, in cooler parts of the Southwest, it is incredibly useful for its ability to grow in shade and fill in blank spots in a garden, adding year-round visual interest and stabilizing loose or eroding soil.

Mimosa dysocarpa – Fabaceae

velvetpod mimosa (gatuño)

4–6′ × 4–6′

full sun

medium water needs

flowers May–September

Field guide: This species forms impenetrable thickets along sloping drainages and rolling hills in Semidesert Grassland and Madrean Evergreen Woodland habitats from 3500 to 6500 feet in the sky islands of Arizona and Sonora.

Description: In the muggy heat of monsoon season, velvetpod mimosa explodes into color with bottlebrush-like sprays of fragrant pink blossoms buzzing with pollinators. A common component of grassland ecosystems in the US–Mexico borderlands, it's notorious for its clothes-ripping thorns but deserves more attention for its value as a landscaping plant. A spreading shrub with an intricate branching pattern and fine deciduous leaflets, this species can be mixed into a pollinator hedge or used as the centerpiece of a garden bed filled with grassland species like tanglehead (*Heteropogon contortus*), sideoats grama (*Bouteloua curtipendula*), and desert spoon (*Dasylirion wheeleri*). It hosts caterpillars of the Reakirt's blue (*Echinargus isola*) and mimosa yellow (*Pyrisitia nise*) butterflies and the enigmatic black witch moth (*Ascalapha odorata*), among others.

Morus microphylla – Moraceae

Texas mulberry (mora)

12–15′ × 10–12′

full sun, partial shade

high water needs

flowers April–September

Field guide: Found between 3500 and 5000 feet in the moist canyons cutting through Interior Chaparral, Semi-desert Grassland, Madrean Evergreen Woodland, and Great Basin Conifer Woodland.

Description: Our native mulberry is much more modest in size than Eurasian mulberries, which become massive trees with spreading canopies that produce long, caterpillar-like fruits. This species reaches not much more than fifteen feet, with delicious fruits that typically don't surpass the size of a fingernail. Regardless, it is a powerhouse bird attractor and serves as a larval food source for the mourning cloak butterfly (*Nymphalis antiopa*). It's a stellar tree for small spaces and thrives in the filtered light found under velvet mesquite (*Neltuma velutina*), Arizona white oak (*Quercus arizonica*), or velvet ash (*Fraxinus velutina*) but will grow nicely as a solo patio tree, where it can be underplanted with evergreen shrubs and wildflowers that balance its winter deciduousness. Native mulberries are generally found in riparian canyons, so this plant will require supplemental irrigation during the warm months. Otherwise, you will find it to be undemanding and highly rewarding to grow.

Mortonia scabrella – Celastraceae

Rio Grande saddlebush

4–6′ × 4′

full sun

low to medium water needs

flowers April–July

Field guide: Oddly distributed, with one large population around the Grand Canyon and the Utah–Arizona border and another hundreds of miles away in southern New Mexico and Arizona. What unites these disparate populations is the affinity of Rio Grande saddlebush for rocky soils heavy in calcium, like sandstone and limestone. Look for this plant between 2500 and 5000 feet.

Description: This upright evergreen shrub with small, rough-textured leaves is ideal for rock gardens and can make a nice informal hedge when planted in groups. The preference of this species for calcareous soils makes it rot-prone in heavy clays. The flowers are small, cream-colored, and not likely to win any beauty contests, but they do a nice job of attracting bees and butterflies. Because it is difficult to cultivate from seeds or cuttings, it is not currently very common in nurseries, but this plant deserves more attention in Southwestern gardens for its ability to thrive in thin, difficult soils.

Parthenium incanum – Asteraceae

mariola (hierba ceniza)

3′ × 3′

full sun

low water needs

flowers June–October

Field guide: This humble desert shrub is found in New Mexico, Arizona, and southwestern Utah on rocky slopes and gravelly wash edges in Chihuahuan, Great Basin, and Sonoran Desertscrub, Interior Chaparral, and Semidesert Grassland habitats between 2500 and 5000 feet.

Description: Despite being a member of the sunflower family, this species has blooms that aren't particularly showy; they consist of a dome-shaped display of small, creamy white, bee-attracting blossoms. The leaves are gray, with soft hairs and a pleasant medicinal fragrance. This scent hints at the fact that mariola is a close relative of guayule (*Parthenium argentatum*), a drought-tolerant shrub cultivated for its commercial value as a source of latex. Mariola is most effective when planted against taller shrubs like jojoba (*Simmondsia chinensis*) and desert hackberry (*Celtis pallida*) or when it's blended into a rocky cactus garden. This plant is unfazed by sun or heat and is suitable for even the harshest planting areas.

Plumbago zeylanica – Plumbaginaceae

leadwort (esplúmbago)

3–5′ × 3–5′

full sun, partial shade

medium water needs

flowers year-round (especially April–November)

Field guide: Thrives in the dappled light along sandy wash edges and canyon margins in Arizona Upland Sonoran Desertscrub and Semidesert Grassland from 2500 to 4500 feet. This subtropical plant is also found in eastern Texas and southern Florida.

Description: Few low-desert plants are well adapted to growing in shade, but leadwort is a definite exception, thriving and blooming prolifically in the shadows cast by buildings or mature trees and giving gardeners a great option for those difficult north-facing exposures. Weather permitting, leadwort can produce a wealth of white blossoms at any time of year, providing a reliable food source for butterflies, bees, and moths; the foliage hosts caterpillars of the dainty marine blue butterfly (*Leptotes marina*). Leaves of this species shift from a rich glossy green to a bruise-purple depending on temperature and sun exposure, with both colors sometimes showing up on a plant at the same time. Extra moisture in summertime will keep leadwort looking happy and ensure a much longer bloom season.

Poliomintha incana – Lamiaceae

frosted mint

3–4′ × 6′+

full sun

low water needs

flowers April–October

Field guide: Found in the Four Corners region and southeastern New Mexico, where it thrives on sandy soils in Chihuahuan Desertscrub, Great Basin Desertscrub, and into Great Basin Conifer Woodland from 4000 to 6000 feet.

Description: This species can withstand the adverse conditions found on sand dunes due to its robust root system and the way its branches spread across the ground, rooting at the nodes to capture and stabilize soil. It is coated in fine hairs that reflect light and give it a frosty appearance, and the foliage exudes an exceedingly pleasant aroma. Gardeners can utilize this species to fill in rock gardens and stabilize slopes and terraces—and periodically harvest some branches for tea if the plant is spreading too far. While frosted mint isn't in a hurry to grow, it is tolerant of both heat and cold, making it an appropriate plant for a wide range of gardens (but not for heavy clay soils). In yards where poor drainage might be an issue, plant it on a mound of dirt and mix in pumice, coarse sand, or even crushed gravel to mimic the dunes this plant is fond of and ensure optimum growth.

Purshia stansburyana – Rosaceae

cliffrose (romerillo cimarrón)

6–10′ × 4–8′

full sun, partial shade

medium water needs

flowers April–September

Field guide: This widespread shrub is found in all the Southwestern states, but it's picky about habitat, preferring thin-soiled rocky slopes and limestone cliffs between 3000 and 8000 feet in the Interior Chaparral, Madrean Evergreen Woodland, Great Basin Conifer Woodland, and Montane Conifer Forest bioregions.

Description: This plant survives on rocky soils by sending down a taproot and numerous smaller feeder roots to access moisture hidden in crags and crevices. While it is similar to Apache plume (*Fallugia paradoxa*), it differs in key characteristics, including its more upright growth habit, smaller flowers, and fewer hairy plumes emanating from the seeds. Evergreen and a source of nearly year-round visual interest from flowers and seeds, cliffrose can be the centerpiece of a garden of smaller shrubs and wildflowers, or it can be set into a hedge with its typical associates like oaks (*Quercus* spp.) and junipers (*Juniperus* spp.). The copious cream-colored blossoms attract native pollinators, including tiny fairy bees (*Perdita* spp.) and various butterflies. Ideal for mid- to high-elevation gardens with adequate soil drainage and space to accommodate this robust and showy shrub.

Quercus turbinella – Fagaceae

scrub oak (encinillo)

5–15′ × 5–15′

full sun, partial shade

low to medium water needs

flowers March–June

Field guide: A widespread species found across the Southwest and especially prominent in Interior Chaparral, Madrean Evergreen Woodland, and Great Basin Conifer Woodland habitats. Found from 2500 to 8000 feet on rocky hillsides, windswept mountains, and gravelly wash bottoms.

Description: This drought-tolerant oak is a versatile landscaping plant capable of handling a wide range of conditions. It tends to be multi-trunked with rough gray bark and small, serrated leaves that can be green or light gray. In extreme exposures with limited moisture, these plants may be small shrubs, but with ample water, scrub oaks can form a dense but low canopy that makes them a fantastic understory tree in a landscape. Be cautious when gardening around oaks, because roots cut below the ground often result in the unsightly dieback of aboveground branches. Like other oak species, this is an excellent wildlife plant that will serve as a nesting habitat and food source for a wide range of birds and mammals.

Rhus aromatica var. *trilobata* – Anacardiaceae

three-leaf sumac (limita)

4–6′ × 5′

full sun, partial shade

low water needs

flowers March–June

Field guide: A widespread shrub found on slopes and canyon margins from 2500 to 7500 feet. This broad elevational range means that three-leaf sumac occurs in a variety of habitats, but it is most common in Interior Chaparral, Madrean Evergreen Woodland, and Great Basin Conifer Woodland.

Description: This plant is appropriate for almost any Southwestern garden and is commonly available at regionally focused plant nurseries. Because it is so widely distributed, three-leaf sumac has many common names that can teach us about its qualities. One name, "skunkbush," refers to the musty-smelling leaves, while "lemonade berry" hints at the tart, tasty berries that can be steeped into a tea, and *hierba de venado* (deer herb) alludes to its association with woodland habitats frequented by deer. Plant this shrub on property edges to form a bird-attracting thicket, and interplant with evergreens such as Emory oak (*Quercus emoryi*), alligator juniper (*Juniperus deppeana*), and Wright's silktassel (*Garrya wrightii*) to mask three-leaf sumac's leafless winter form.

Rhus microphylla – Anacardiaceae

littleleaf sumac (lima de la sierra)

6′ × 6′

full sun, partial shade

medium water needs

flowers March–May

Field guide: A species well adapted to stony soils and rocky outcrops and common in open rolling hills and canyon edges in southeastern Arizona and across New Mexico. This plant is found in Chihuahuan Desertscrub as well as Semidesert Grassland, Great Basin Grassland, and into Madrean Evergreen Woodland from 3500 to 6500 feet.

Description: Easily one of the most drought-tolerant shrubs for mid-elevation gardens, littleleaf sumac is also an excellent wildlife plant, attracting a variety of pollinators with Christmas-ornament-like red berries that are a favored food source for myriad bird species. Stout and wide, littleleaf sumac often occurs in dense masses, and this is a habit that can be recreated in a landscape to edge a garden bed or line a driveway.

Rhus ovata – Anacardiaceae

sugar sumac (lentisco)

15′ × 15′

full sun

low water needs

flowers February–May

Field guide: The range of sugar sumac runs northwest to southeast across the Interior Chaparral plant communities found in the mountains of central Arizona, where it often grows on exposed rocky outcrops or mixed into Great Basin Conifer Woodland from 3500 to 6000 feet.

Description: This plant's drought and heat tolerance makes it appropriate even for low-desert gardens, where its glossy leaves add a lovely texture to the landscape. Because of its large size and evergreen habit, sugar sumac serves well as a hedge or screen, and its density makes it a favored nesting site for a variety of birds. In spring, it's covered in red and white blossoms followed by red fruits covered in crystals of malic acid that give them a tart flavor appealing to birds and people. In the sky islands along the boundary with Mexico, this species is replaced by evergreen sumac (*Rhus virens* var. *choriophylla*).

Ribes aureum – Grossulariaceae

golden currant

4′ × 4′

full sun, partial shade, shade

medium water needs

flowers March–July

Field guide: A resident of Riparian Corridors, where it is found in moist meadows and shady stream edges from 4000 to 7000 feet.

Description: This species forms small patches of woody shoots arising from a common root system, making it an ideal choice for planting along a fence or wall. The prolific flowers are vibrant yellow tubes with crimson throats; they exude a pleasing aroma that guides in pollinators like hummingbirds and butterflies. The blooms are followed by burgundy berries with a tart flavor that is beloved by birds and make a fine jam. In fall, the leaves will turn red before dropping with winter cold. At lower elevations, golden currant requires ample water and afternoon shade, but it can be planted in the open in upland gardens. This plant does well around a basin and makes a good informal hedge along the edges of a vegetable garden.

Ribes pinetorum – Grossulariaceae

orange gooseberry

4′ × 4–6′

full sun, partial shade

medium water needs

flowers April–September

Field guide: Grows in moist meadows and along stream margins in Montane and Subalpine Conifer Forest habitats from 7000 to 10,000 feet.

Description: With aggressively spiny fruit and intimidating thorns along the stems, orange gooseberry is not a species you will soon forget after first contact. This shrub is an ideal selection for partially shaded areas in high-elevation gardens and can be planted under pinyon pine (*Pinus edulis*), bigtooth maple (*Acer grandidentatum*), and white fir (*Abies concolor*). The fruits, though armed, are fairly tasty if eaten carefully, and the scarlet flowers are attractive to people and pollinators alike. Like other *Ribes* species, orange gooseberry is a host for the hoary comma butterfly (*Polygonia gracilis*) and western tent caterpillar moth (*Malacosoma californicum*) and attracts a diverse array of pollinators when in flower through spring and summer.

Robinia neomexicana – Fabaceae

New Mexico locust (uña de gato)

15–25′ × 15′

full sun, partial shade

medium to high water needs

flowers May–August

Field guide: Found across the Southwest, where it is most common in Great Basin Conifer Woodland and Montane Conifer Forest habitats. This species is an understory shrub that occurs along canyon edges and forest clearings from 5000 to 8000 feet.

Description: Cold-tolerant, easy to cultivate, and sporting masses of showy flowers, this large shrub has many qualities that recommend its use in Southwestern landscapes. It tends to appear after fires or forest clearing because it can quickly send up stems from its root system to form thickets in open ground. The thorny stems are cloaked in oval-shaped compound leaflets that turn a saffron-yellow before dropping at the onset of winter frosts. The foliage plays host to a diverse assemblage of butterflies and moths, including the cecrops eyed silkmoth (*Automeris cecrops*) along with funereal duskywing (*Erynnis funeralis*) and silver-spotted skipper (*Epargyreus clarus*) butterflies. Plant where the root sprouts will be a benefit, not a nuisance, as the plant forms a living fence that will burst into a profusion of pink, pollinator-attracting blossoms in spring.

Rosa woodsii – Rosaceae

Woods' rose (rosa)

4–6′ × 4–6′

partial shade, shade

high water needs

flowers June–September

Field guide: Grows in moist meadows and stream banks in Great Basin Conifer Woodland and Montane Conifer Forest habitats from 5500 to 9000 feet in mountains across the Southwest.

Description: This charming wild rose forms spiny brambles arising from a spreading root system that knits together loose soil. The vivid pink flowers appear in summer, and the ruby-red, grape-sized rose hips are seedy but edible, with a tart but pleasant flavor. While Woods' rose can survive in desert gardens, it rarely thrives without shade and considerable water investment. However, in the upland and mountain Southwest, this species makes a good low-maintenance hedge or understory planting that will attract wildlife ranging from gall-forming moths to nectar-feeding butterflies and fruit-eating birds. Because it spreads by the roots, Woods' rose can be competitive with other plants, so give it plenty of room to stretch its legs. Consider using it on the edge of a drainage or swale to help hold the banks in place.

Sambucus cerulea (syn. *Sambucus nigra* subsp. *cerulea*) – Viburnaceae

blue elderberry (tápiro)

15–20′ × 15–20′

full sun, partial shade

medium to high water needs

flowers May–August

Field guide: This species covers a range stretching from Mexico to Canada, where it is found in wetlands and Riparian Corridors from 2500 to 5000 feet.

Description: Blue elderberry is a large deciduous shrub with flat-topped clusters of small white flowers that attract bees and butterflies. These flowers are followed by bunches of small purple berries that are relished by birds and mammals but poisonous to humans until cooked to neutralize the toxins. With ample irrigation, elderberries can grow to become large hedges or screens, but if water is withheld they will suffer. Elderberry is an excellent addition to a native bird garden with pale wolfberry (*Lycium pallidum*), netleaf hackberry (*Celtis reticulata*), and Texas mulberry (*Morus microphylla*). Plant in a basin and watch as seedlings sprout in irrigated parts of the yard where birds deposit the berries.

Sarcomphalus obtusifolius (syn *Ziziphus obtusifolia*; *Condaliopsis divaricata*) – Rhamnaceae

graythorn (bachata)

6–9′ × 6–9′

full sun, partial shade

low water needs

flowers May–September

Field guide: Grows in mesquite thickets and along dry washes in the Sonoran, Mojave, and Chihuahuan deserts into Semidesert Grassland and Interior Chaparral habitats from 1000 to 5000 feet.

Description: Graythorn is an often-overlooked and unappreciated member of desert plant communities in the Southwest, where its ashy gray bark, prodigious spines, and lack of showy blossoms prevent people from noticing its immense value to wildlife. Caterpillars of the Pyrrha's prominent moth (*Cargida pyrrha*) eat the leaves, bees and butterflies pollinate the small flowers, and birds relish the dark berries and the habitat provided by the thorny branches. Graythorn is part of the velvet mesquite (*Neltuma velutina*) and wolfberry (*Lycium* spp.) community that formerly graced washes and rivers across the desert Southwest but has largely disappeared because of human disturbance. Plant it with the above-mentioned species along a fence or in an alleyway to create a wildlife haven and an impenetrable hedge.

Simmondsia chinensis – Simmondsiaceae

jojoba (jojoba)

4–6′ × 6–8′

full sun, partial shade

low water needs

flowers February–May

Field guide: A common component of Lower Colorado River Sonoran Desertscrub and Arizona Upland Sonoran Desertscrub plant communities, where it grows from 1500 to 5000 feet along gravelly arroyos and rocky slopes.

Description: Even if you've never seen this plant, chances are you've felt it on your skin. The oil pressed from the fruits is nearly ubiquitous in skincare and hair products, making this shrub one of the most economically important native plants in the desert Southwest. Often grows in association with foothill paloverde (*Parkinsonia microphylla*) and saguaro (*Carnegiea gigantea*), and this trifecta can be transposed into gardens in central and southern Arizona. The flowers are nondescript, with pollen- and seed-bearing blooms borne on separate individuals. The pollen is wind-dispersed, but bees sometimes visit the flowers. Leathery oval leaves are oriented upright on the stems to minimize sun exposure. Jojoba's evergreen habit and large size make it a good choice for an informal hedge along a fence or property line, where the dense branches will serve as perches for desert birds like cactus wrens (*Campylorhynchus brunneicapillus*) and curve-billed thrashers (*Toxostoma curvirostre*).

Tecoma stans – Bignoniaceae

yellow bells (lluvia de oro)

5–7′ × 5′

full sun

low to medium water needs

flowers May–November

Field guide: Flourishes on rocky hillsides, boulder fields, and canyon edges along the US–Mexico boundary, in Semi-desert Grassland and Madrean Evergreen Woodland habitats from 3000 to 5000 feet.

Description: This species forms patches of plants that launch into displays of golden-yellow bell-shaped blossoms that can last from spring until the first frost. It's increasingly popular in desert landscapes, as evidenced by the numerous named cultivars available at nurseries. Gardeners are advised to stick with our native variety (var. *angustata*), which has narrow leaves and exhibits greater drought and frost tolerance than the big-leafed nursery cultivars while still providing ample food for the caterpillars of calleta silkmoth (*Eupackardia calleta*) and various species of sphinx moths (*Manduca* spp.). Give it a prominent spot in your landscape as part of a bed of flowering shrubs, an element in a naturalistic Madrean Evergreen Woodland planting, or right by a wall or entryway where visitors can appreciate its lush, tropical aesthetic and prolific blooms.

Vachellia constricta – Fabaceae

whitethorn acacia (huizache)

6–12′ × 6–12′

full sun

low water needs

flowers March–September

Field guide: This species is fond of sandy washes and rocky slopes from 2000 to 5000 feet in the Sonoran and Chihuahuan deserts.

Description: A shrub so common in the deserts of the Southwest that it can be easy to overlook. But when the yellow puffball flowers appear in spring, their pleasant honey-like aroma drifts on the breeze, and their lime-green leaves provide a pop of color against rocky hillsides and wash edges. Ideal for a low-water hedge with desert hackberry (*Celtis pallida*) and Berlandier's wolfberry (*Lycium berlandieri*) that will be impassible to people and irresistible to pollinators and birds. The feathery foliage is readily eaten by black witch (*Ascalapha odorata*), merry melipotis (*Melipotis jucunda*), and mesquite stinger (*Norape tener*) moths. Dense and multibranched, this is easily differentiated from other acacias by its bushy form and pairs of long white thorns, except where its range overlaps with viscid acacia (*Vachellia vernicosa*). That similar species differs in being lankier and having resinous leaves. For a drought-tolerant desert landscape shrub, whitethorn acacia is a top choice.

Vauquelinia californica – Rosaceae

Arizona rosewood (árbol prieto) 

10–15′ × 8–10′

full sun, partial shade

low water needs

flowers May–July

Field guide: Grows on rocky slopes and is commonly seen growing straight from cliff faces, especially where limestone is present. This plant is found in Semidesert Grassland and Madrean Evergreen Woodland habitats from 2500 to 5000 feet.

Description: This excellent screening or specimen shrub makes an ideal replacement for non-native oleanders (*Nerium oleander*) because of its dense evergreen growth and lack of the toxic sap that can make oleanders a danger to pets and children. These plants do best in full sun, where they will develop upright branches with dark, lance-shaped leaves, topped with clusters of small, five-petaled white flowers. Plant in well-draining soil to prevent rot and improve its modest growth rate. Once established, Arizona rosewood is extremely tough, and a few deep waterings during the dry season should be sufficient to maintain this plant, even in the scorching urban environments of Phoenix and Tucson.

Perennials

For many people, purchasing, growing, and planting wildflowers is their favorite part of gardening. Wildflowers are often what we think of when we talk about gardening: rows of showy blooms buzzing with pollinator activity and stopping neighbors in their tracks as they pass. With your trees and shrubs in place, you can use wildflowers to tie the yard together, knowing that these smaller plants will reach their prime as slower-growing specimens are just getting their roots under them. There are a variety of design elements to consider when selecting wildflowers. For instance, what colors will you blend for maximum floral effect? Which pollinators are you hoping to see in your yard? How can you select species with differing bloom times to provide food for pollinators and a feast for the eyes throughout the year? Considering these factors in advance will allow you to assemble a fantastic array of perennial flowers that will quickly turn your garden into a pollinator paradise. Keep in mind that many of these plants will be short-lived species that either fade out as large plants become established or will reseed themselves in abundance, thereby establishing self-sustaining populations that will be with you for years. Get creative, and don't be afraid to experiment. Gardening with wildflowers is a rewarding experience with none of the delayed gratification that comes with trees, shrubs, and cacti.

Achillea millefolium – Asteraceae

yarrow (canfor)

1.5′ × 1′

full sun, partial shade, shade

medium to high water needs

flowers June–September

Field guide: A fascinating species whose range circumnavigates the globe and features in the pharmacopeia and cultural practices of innumerable societies. In the Southwest, it is found between 5500 and 11,500 feet in Riparian Corridors, and Montane Conifer Forest, and Subalpine Conifer Forest habitats.

Description: This plant's wide distribution and ease of growth have led to the development of many nursery cultivars with various flower colors. Even so, the white blooms of the wild plant are a fantastic complement to any pollinator, woodland, or native perennial garden. Yarrow prefers soil with some organic matter and will do well even in fairly deep shade. Under ideal conditions, it spreads readily by its root system; this weedy habit can be a great benefit if it's planted onto a terraced slope to prevent erosion or to fill in a rock-edged bed that has shade from nearby trees. The rosette of ferny leaves pushes up narrow flower stalks, which are topped by flat clusters of cream-colored blossoms that attract a diverse array of bee species.

Argemone pleiacantha – Papaveraceae

southwestern pricklypoppy (chicalote)

3′ × 1′

full sun, partial shade

medium water needs

flowers April–September

Field guide: Look for this plant in washes and disturbed sites between 2500 and 7500 feet. It's widely distributed but most often found in Semidesert Grassland, Madrean Evergreen Woodland, and Great Basin Conifer Woodland habitats and into Interior Chaparral and the lower reaches of Montane Conifer Forest. Mostly found in Arizona and eastern New Mexico, often in roadside ditches.

Description: This common roadside wildflower has a bloom that resembles a fried egg and gray-blue foliage covered in nettle-like thorns. It's unmistakable, even when you are passing it at sixty miles per hour. Because it prefers disturbed areas, pricklypoppy can fill in open lots while longer-lived trees and shrubs establish. It also makes a good planting for alleys, driveway edges, and unfenced areas where its prickly foliage and latex sap will deter browsing critters that might devour other plants. The large blossoms with their milky-white petals, mass of yellow stamens, and sticky pistil are favored by a wide variety of insects, from bumblebees and butterflies to beetles that will gobble up pollen and munch around the margins of the flower. An abundant seed producer, pricklypoppy will often volunteer, especially in sandy soils.

Aquilegia chrysantha – Ranunculaceae

golden columbine (aquileña)

2′ × 2′

full sun, partial shade

high water needs

flowers March–September

Field guide: This species grows across a very wide elevational range—from 3000 to 11,000 feet—but is always associated with Riparian Corridors, seeps, and moist meadows.

Description: Golden columbine enlivens gardens with neat little clumps of leaves, from which thin stalks rise another foot or so to present the elegant flowers. The flowers start as almond-shaped buds, followed by narrow blossoms with graceful spurs, and ultimately become glorious golden blooms that bob in the breeze. Columbine's long spurs hold the plant's nectar, enticing hawk moths (*Sphingidae* spp.) and occasionally hummingbirds, who use their long tongues to reach into the bloom. This species may be grown at nearly any elevation but requires shade in lower-elevation gardens, where it often dies back and looks somewhat ratty in the early summer. This species wants reliable moisture at any elevation and is best suited to basins, small ponds, and gray water–fed gardens.

Aquilegia desertorum – Ranunculaceae

desert columbine

2′ × 2′

full sun, partial shade

medium to high water needs

flowers May–October

Field guide: This showy wildflower is found around Riparian Corridors, meadows, and seeps primarily in Madrean Evergreen Woodland, Montane Conifer Forest, and Subalpine Conifer Forest between 4000 and 10,000 feet. Look for it on the rocky outcrops and patches of limestone that it favors.

Description: More drought-tolerant than golden columbine (*Aquilegia chrysantha*), desert columbine grows directly out of bedrock, making it an ideal choice for rock gardens and terraces. With its ability to handle low-light conditions, this species can be used in parts of a landscape with overhanging trees or shade cast by walls, where it can be a showy and long-blooming companion for other shade-tolerant plants such as showy fleabane (*Erigeron speciosus*), tasselflower brickellbush (*Brickellia grandiflora*), and ricegrass (*Achnatherum hymenoides*). The red and gold blooms of desert columbine are particularly attractive to hummingbirds, and the long lifespan of this plant means it will continue to delight in high-elevation landscapes year after year.

Artemisia frigida – Asteraceae

fringed sagebrush

1′ × 2′

full sun, partial shade

low water needs

flowers July–October

Field guide: Found between 5500 and 10,000 feet on rocky slopes and flats, from Great Basin Desertscrub and Great Basin Conifer Woodland through Montane Conifer Forest to meadows in Subalpine Conifer Forest habitats. The range extends north to Canada, and this species can even be found in Mongolia and Siberia.

Description: Despite its nondescript wind-pollinated flowers, this plant is a worthy inclusion in mid- to high-elevation gardens. Think of it as the rug that ties the whole room together, with lacy silver foliage that contributes year-round visual interest to a garden. Aside from this mounding subshrub's aesthetic appeal, its roots also bind up loose soil and prevent erosion on slopes or washouts during heavy rains. Additionally, the leaves are a larval food source for a variety of moth and butterfly species, so although the blooms are wind-pollinated, it's an insect-friendly plant that will encourage pollinators to visit. Plant alongside darker green species like damianita (*Chrysactinia mexicana*), Dakota mock vervain (*Glandularia bipinnatifida*), and fernbush (*Chamaebatiaria millefolium*) for a lovely color contrast.

Asclepias subverticillata – Apocynaceae

horsetail milkweed (hierba lechosa)

3′+ × 3′+ (spreads by runners)

full sun, partial shade

medium to high water needs

flowers May–October

Field guide: Found between 3000 and 8000 feet in grassy marshes, roadside ditches, and moist meadows in all Southwestern habitats except the Sonoran and Mojave deserts and Subalpine Conifer Forest.

Description: The pollinator-plant relationship that comes most readily to mind for the average person is between milkweeds and monarchs (*Danaus plexippus*). Monarchs are highly desirable in a landscape, and luckily for gardeners, their host plants are some of our loveliest wildflowers. Of the Southwest's more than thirty milkweed species, this is one of the best for feeding monarch and queen (*Danaus gilippus*) butterflies. It spreads by seeds and underground runners to form patches of upright stems with thin leaves and clusters of small white blossoms that attract butterflies, bees, and wasps. Butterfly caterpillars have evolved to process the toxic sap of this plant, and other insects, such as bright orange oleander aphids (*Aphis nerii*) and red milkweed beetles (*Tetraopes tetrophthalmus*), also eat horsetail milkweed. These insects do minimal damage to the plant and bring insect predators like lacewings, spiders, and syrphid flies. Horsetail milkweed spreads quickly, ensuring that there will be enough to go around, and this one species can support a whole ecosystem in a garden.

Asclepias tuberosa – Apocynaceae

butterfly milkweed (inmortal)

2′ × 2′

full sun, partial shade

low water needs

flowers May–September

Field guide: Widespread across North America, but in the Southwest primarily found in Riparian Corridors and woodland clearings in Semidesert Grassland, Madrean Evergreen Woodland, Great Basin Conifer Woodland, and Montane Conifer Forest habitats between 3000 and 8000 feet.

Description: Of all our native milkweeds, this species may be the showiest in a garden. Its bright orange or golden-yellow flowers and neat arrangement of dark green, lance-shaped leaves would make this a popular landscaping plant even without its immense value to a host of pollinators, including the ever-popular monarch and queen butterflies (*Danaus plexippus, D. gilippus*). Butterfly milkweed isn't fussy in cultivation, but its sizeable tuberous root means this species is best suited to sandy soil and may languish, or even rot, in heavy clays. Plant in a well-draining pollinator bed with other butterfly favorites like mistflower (*Conoclinium dissectum*), Goodding's mock vervain (*Glandularia gooddingii*), and fall tansyaster (*Dieteria asteroides*).

Baileya multiradiata – Asteraceae

desert marigold (hierba amarilla)

2′ × 2′

full sun

low water needs

flowers year-round

Field guide: Thrives in a variety of habitats but is most at home in the Sonoran, Mojave, and Chihuahuan deserts between 100 and 5000 feet, where it is often found in sandy washes and roadside ditches or on gravelly flats where it can form extensive stands.

Description: Already one of the most popular flowers for low-water landscaping, this reliable garden plant is equally suited to a wildflower patch or a cactus garden. The yellow blossoms persist nearly year-round and provide a good nectar source for bees and butterflies as well as beetles, who dine on both the pollen and petals. Also hosts a specialized moth called the desert marigold moth (*Schinia miniana*) that lays its eggs on the flowers, where the larvae have a readily available food source in the petals. A handful of initial plantings can reseed and give rise to many subsequent individuals that will contribute golden flowers and distinctive powdery-white foliage to your yard. Desert marigold isn't fussy about its conditions and seems to grow as happily in the cracked pavement of a back alley as in a fertilized and meticulously maintained flower bed.

Berlandiera lyrata – Asteraceae

chocolate flower (coronilla)

2′ × 2′

full sun, partial shade

low to medium water needs

flowers March–October

Field guide: You are most likely to encounter this plant in the sandy flats, rolling hills, and limestone outcrops of Chihuahuan Desertscrub and Great Basin Desertscrub habitats and into Great Basin Grassland between 2500 and 7000 feet. Widespread in New Mexico, this species occurs in the southeastern corner of Arizona but is common in cultivation around the Southwest.

Description: This species is easily recognizable by its fragrance—a strong, malty aroma reminiscent of melting chocolate—and its pollen-bearing stamens, which even taste a bit like the real thing. The lobed leaves are topped by blooms with daisy-like yellow rays and velvety maroon disc flowers. Chocolate flower can reseed and spread in a garden, particularly where light shade and supplemental irrigation are present. Include this species in a butterfly or bee garden with other native grasses and wildflowers for a meadow planting that attracts pollinators and stimulates the senses.

Brickellia grandiflora – Asteraceae

tasselflower brickellbush

3′ × 3′+

full sun, partial shade

medium to high water needs

flowers July–October

Field guide: A species of woodlands and forests in the mountains of the Southwest, where it can be found on duff-covered slopes under conifers and oaks from 5000 to 10,000 feet.

Description: The rough, triangle-shaped leaves of this plant are borne on long petioles, and the green-yellow flowers occur in clusters that nod as if the weight of the blossoms is too much for the plant to bear. The blooms are modest in color but frequently play host to visiting butterflies and are also eaten by caterpillars of flower moths (*Schinia* spp.); the foliage is a larval food source for the giant northern flag moth (*Dysschema howardi*). In high-elevation gardens where a large tree dominates the yard space, tasselflower brickellbush can be planted with other shade-tolerant flowers and shrubs and watered through the warm months to keep it lush and encourage flowers.

Callirhoe involucrata – Malvaceae

purple poppymallow

1′ × 2′

full sun, partial shade

medium water needs

flowers year-round (especially March–September)

Field guide: This common flower of the Great Plains just makes it into the Southwest in southern Colorado and northeastern New Mexico, where it can be found in sandy plains and roadside ditches from 4000 to 7000 feet.

Description: One of the Southwest's showiest mallows, this is an excellent addition to high-elevation flower beds and pollinator gardens, where it combines well with the white flowers of yarrow (*Achillea millefolium*), the lacy silver foliage of fringed sagebrush (*Artemisia frigida*), and the fine blades of ricegrass (*Achnatherum hymenoides*). The magenta, bowl-shaped blossoms are a draw for native bees, and the mat of trailing foliage is a larval food source for gray hairstreak (*Strymon melinus*) and painted lady (*Vanessa cardui*) butterflies. At lower elevations, purple poppymallow will benefit from shade, and at any elevation a well-draining soil will ensure robust growth. Try planting purple poppymallow in an elevated rock garden or raised planter where it can spread and cascade over the edges.

Calylophus hartwegii – Onagraceae

Hartweg's sundrops

1′ × 2′

full sun, partial shade

low water needs

flowers March–September

Field guide: Look for sundrops on rocky outcroppings, cliff edges, and sandy plains in grassland and woodland habitats from 3500 to 7000 feet.

Description: Sundrops are one of our cheeriest wildflowers, with trailing masses of stems that sport evening-blooming yellow flowers. Like other members of the primrose family, the sundrop is a pollinator powerhouse that attracts bees and butterflies as well as night-flying pollinators like the white-lined sphinx moth (*Hyles lineata*). A fantastic moth garden can be made by planting sundrops with sacred datura (*Datura wrightii*), Colorado four o'clock (*Mirabilis multiflora*), and tufted evening primrose (*Oenothera cespitosa*). Planting this species will give you a front-row seat to the evening visitation of our most enigmatic pollinators.

Campanula rotundifolia – Campanulaceae

bluebell bellflower

1.5′ × 1.5′

partial shade

medium to high water needs

flowers May–October

Field guide: This low-growing wildflower has a global distribution that makes it a common feature of temperate forests and subalpine habitats around the world. In the Southwest, it is restricted to mountains between 8000 and 12,500 feet in forest clearing and subalpine meadows.

Description: As the plant's name suggests, the flowers of this species are bell-shaped with a charming lilac color. They are visited by a large diversity of bees, with small bees climbing to the base of the blossom to access nectar and larger bees foraging for easier-to-reach pollen. This is an ideal flower for shady high-elevation gardens, where the long flowering season will ensure dainty and delightful blooms for several months. Though individual plants are short-lived, the masses of tiny seeds ensure that volunteers are forthcoming, which can create a self-perpetuating population in your garden.

Commelina erecta – Commelinaceae

whitemouth dayflower (hierba del pollo)

1′ × 3′+ (spreads by rhizomes)

full sun, partial shade

medium water needs

flowers May–November

Field guide: This Riparian Corridor species is fond of wet meadows and stream edges in southeastern Arizona and into New Mexico and southeastern Colorado, between 3000 and 7500 feet.

Description: A prolifically spreading plant that will grow under a tree or fill in a basin beautifully. The grass-like leaves of this decorative species rise from a fleshy root, and the stem tips are topped with a beak-like bract (known as a spathe) from which the flowers emerge. The blooms consist of two rich-blue rounded petals that resemble mouse ears around a white bottom petal. The flowers are attractive to a plethora of bees; the stems and leaves are a food source for leaf-eater beetles in the family *Chrysomelidae* and foraging desert tortoises (*Gopherus* spp.). The seeds are relished by mourning doves (*Zenaida macroura*) along with other backyard birds. Because of its ability to aggressively spread, whitemouth dayflower is best reserved for spaces under trees or large shrubs with which it won't compete.

Conoclinium dissectum – Asteraceae

mistflower

2′ × 4′+ (spreads by rhizomes)

full sun, partial shade

medium water needs

flowers March–November

Field guide: Found from 3000 to 5000 feet in sandy washes and rocky slopes in Semidesert Grassland and Madrean Evergreen Woodland habitats in southern Arizona and New Mexico.

Description: Even an insect-averse gardener wants butterflies in their landscape, and few plants will attract them like mistflower. Spreading by underground runners, a few mistflower plantings can quickly fill a bed or large container with a dense cover of dissected leaves and thin stalks bearing sky-blue tasseled blossoms that swarm with butterflies of every size and color. The relationship of mistflower with queen butterflies (*Danaus gilippus*) is fascinating. Most of the queen butterflies you will see on your mistflower are males, identifiable by a pair of black spots on the hind wings. These bachelor butterflies are not only taking in nectar but also a compound called *intermedin*, which is poisonous to browsers but safe for these butterflies (they convert it into a pheromone to attract mates). The male transfers some of this compound to their mate as a "nuptial gift" that will make her eggs unappealing to predators. All of this means that mistflower not only attracts hordes of butterflies, but also ensures the next generation of queen butterflies to delight you.

Dalea albiflora – Fabaceae

whiteflower prairie clover

3′ × 2′

full sun, partial shade

low water needs

flowers March–October

Field guide: Grows in Interior Chaparral, Semidesert Grassland, and Madrean Evergreen Woodland habitats from 3500 to 7500 feet on rocky slopes and sandy washes.

Description: Among our finest butterfly plants in the Southwest are members of the genus *Dalea*, and this is one of the best species in the genus for landscaping. This upright subshrub has lacy foliage that exudes a pleasant aroma and serves as a food source for several butterfly caterpillars, such as those of the dainty Reakirt's blue (*Echinargus isola*) and vibrant southern dogface (*Zerene cesonia*). The inflorescence is cone-shaped with spikes of small, milky-white flowers that are irresistible to bees and butterflies. Requires good drainage but is cold- and heat-tolerant and will handle full sun against a hot garden wall, or the cool shade of an Emory oak (*Quercus emoryi*) interplanted with three-leaf sumac (*Rhus aromatica* var. *trilobata*) and plains lovegrass (*Eragrostis intermedia*).

Datura wrightii – Solanaceae

sacred datura (toloache grande)

3′ × 4′

full sun, partial shade

low water needs

flowers April–October

Field guide: Revels in disturbed areas and grows across a wide range of bioregions through much of the Southwest below 6000 feet, where it does best in sandy, well-draining soils.

Description: This charismatic and beguiling wildflower has massive, bell-shaped white blooms that appear in the evening and reflect the moonlight to attract night-flying pollinators—particularly the tobacco hornworm moth (*Manduca sexta*), which pollinates the blossoms and lays its eggs on the leaves. The blooms wilt and close in the morning to be followed by spine-covered orbs that have earned this species the colorful sobriquet "thornapple." The seeds of datura have an elaiosome, a packet of lipids and proteins that attracts small armies of ants, who disperse the seeds under shrubs or around their colonies. During cold snaps, the aboveground portion of the plant will die, retreating to a substantial tuber under the soil and leaving a spidery skeleton behind. The unpleasant psychedelic and physiological effects of this plant should be a consideration for those with pets or small children that like to browse.

Dicliptera resupinata – Acanthaceae

Arizona foldwing (ramoneada flor morada)

2′ × 3′

full sun, partial shade

low water needs

flowers April–October

Field guide: Found in the US–Mexico borderlands from 2500 to 6000 feet, in Arizona Upland Sonoran Desertscrub, Semidesert Grassland, and Madrean Evergreen Woodland habitats in sandy washes or up rocky slopes.

Description: This prolifically seeding subshrub forms clumps two feet high by three wide, and it is particularly well-suited to basins or sandy path edges. As the common name "foldwing" implies, its two-lipped flowers emerge from a folded, heart-shaped bract. This bract also holds the seeds, which quickly fall and readily germinate, often creating dense patches of seedlings in a garden bed. Butterflies and bees visit the flowers, and the foliage is a larval food source for Texan crescent butterflies (*Anthanassa texana*).

Dieteria asteroides – Asteraceae

fall tansyaster

3′ × 2′

full sun, partial shade

low water needs

flowers March–November

Field guide: This plant's range takes it from nearly sea level in Arizona Upland Sonoran Desertscrub and Lower Colorado River Sonoran Desertscrub to 8000 feet in Montane Conifer Forest. It prefers sandy or gravelly soils and may be found on flats or slopes.

Description: A member of the sunflower family, this plant sports cheery purple ray petals and bright golden disc flowers, borne at the top of lanky stems rising two to three feet from the ground. As with other asters, the blooms are ideal landing platforms for pollinators like bees and butterflies, and the hairy foliage sustains the caterpillars of brown-hooded owlet (*Cucullia convexipennis*) and darker-spotted straw (*Heliothis phloxiphaga*) moths. The broad range and temperature tolerance of this perennial species make it appropriate for almost any Southwestern garden, though in extreme heat or cold, this plant may die in either winter or summer. Fall tansyaster produces an abundance of seed that readily germinates, and one plant will easily yield several more. This is a versatile and attractive wildflower that no garden should be without.

Epilobium canum
(syn. *Zauschneria californica*) – Onagraceae

hummingbird trumpet (balsamea)

2′ × 4′+

full sun, partial shade

low to medium water needs

flowers May–October

Field guide: Almost always associated with sandy washes and gravelly canyon margins from 2500 to 7000 feet, this species is especially common in Semidesert Grassland habitats, but its range extends into the Sonoran Desert and up to Madrean Evergreen Woodland as well.

Description: Already widespread in the nursery industry, this plant needs little introduction; it is stunning, adaptable, and one of the Southwest's premier hummingbird plants. It lends itself well to a basin or gray-water planting where the spreading root system will hold together the soil, and the extra moisture will ensure blooms through summer and well into fall, when other nectar resources begin to disappear. The long, red, tubular blossoms are ideally suited for hummingbird pollination, but butterflies will happily sink their proboscises in. Like other members of the primrose family, this plant's leaves will be browsed by white-lined (*Hyles lineata*) and Clark's sphinx (*Proserpinus clarkiae*) moth caterpillars. In colder climates it may freeze to the ground in winter, but it will pop back up from the root system in spring. Even in warmer climates, it can be helpful to cut the plant back periodically to encourage a lush, bushy habit with lots of blooms.

Erigeron speciosus – Asteraceae

showy fleabane

3′ × 3′

full sun, partial shade

medium water needs

flowers July–October

Field guide: Grows in shaded forest edges and cool thickets from Madrean Evergreen Woodland and Great Basin Conifer Woodland into Montane Conifer Forest from 6000 to 9500 feet.

Description: *Speciosus* means "showy" or "beautiful," and it's nice to see that there is still some truth in advertising. Typically found in the shade of overhanging trees, this upright wildflower is topped by white-to-lavender blossoms with bright yellow bullseyes in the center. As with most asters, the flowers are popular with butterflies, bees, and even pollinating flies, and like other fleabanes, this plant is a larval food source for various moths and butterflies. Try showy fleabane in a shady spot on the north side of a structure, or plant it underneath a grove of Gambel oak or silverleaf oak (*Quercus gambelii, Q. hypoleucoides*) and pinyon pine (*Pinus edulis*) alongside three-nerve goldenrod (*Solidago velutina*). In lower elevations, the more drought- and heat-adapted spreading fleabane (*Erigeron divergens*) should be substituted for this species.

Eriogonum umbellatum – Polygonaceae

sulfur buckwheat (alforfón)

3″–9″ × 3–4′

full sun, part shade

low water needs

flowers May–October

Field guide: Grows from 4000 to 9000 feet on rocky slopes, open plains, and woodland clearings from Great Basin Grassland through Montane Conifer Forest.

Description: Where fire or fallen trees have created clearings in pine-oak woodlands, sulfur buckwheat sends out its sprawling stems. Thin petioles hold spatula-shaped leaves covered in a cobwebby fuzz, and from these leaves rise short stems topped by clusters of tiny yellow flowers. This species is best used in high-elevation gardens, where it will tolerate extreme cold and sandy soils. Plant at the base of boulders with pine (*Pinus* spp.) and sagebrush (*Artemisia* spp.) nearby or on the edge of a rock garden where the stems will cascade down. As with all buckwheats, the flowers are a pollinator magnet, especially favored by dainty hairstreak and blue butterflies (*Lycaenidae* spp.). This is one of the most reliable groundcovers for cold gardens in the Southwest.

Eryngium heterophyllum – Apiaceae

Wright's eryngo (hierba del sapo)

1.5′ × 1.5′

full sun, partial shade

low water needs

flowers July–October

Field guide: Typically found in summer in Semidesert Grassland and Madrean Evergreen Woodland habitats, from 4000 to 6500 feet in open meadows and canyon edges.

Description: Wright's eryngo is an unusual flower with a ghostly pale color and a skeletal frame that terminates in pointy bracts subtending an orb of tiny blossoms. This odd member of the carrot family sprouts from a taproot following summer rains, its light color contrasting with the dark green of grasses and oaks. Bees, flies, and butterflies will all visit the clusters of flowers, making this interesting and unusual pollinator plant a good choice for a short-grass prairie or wildflower planting. The flower stalks are particularly lovely in dried floral arrangements.

Erysimum capitatum – Brassicaceae

western wallflower

2–3′ × 2′

full sun, partial shade

medium water needs

flowers March–September

Field guide: This wildflower grows in a diversity of habitats and can be found across the Southwest from 3000 to 9500 feet, on dunes in Great Basin Grassland to slopes and meadows in Montane Conifer Forest.

Description: Western wallflower delights with its windmills of orange or yellow petals that stand out as bright beacons against the rich greens of mountain meadows or the stark white of sand dunes. Though this species is short-lived, the thin seed-bearing capsules dry and split to drop innumerable small seeds into the soil, making this a reliable addition to a self-sustaining wildflower bed. A pollinator generalist, it attracts butterflies, bees, and even ants, while the seeds are eaten by insects and small birds. A member of the butterfly-favorite mustard family, wallflower provides food for the larvae of several species, including checkered white (*Pontia protodice*) and desert marble (*Euchloe lotta*) butterflies.

Erythranthe cardinalis – Phrymaceae

scarlet monkeyflower

2′ × 3′+

partial shade

high water needs

flowers May–October

Field guide: This Riparian Corridor plant can be found along stream edges, seeps, and canyon walls from 2000 to 8000 feet.

Description: Scarlet monkeyflower has brilliant butterfly-attracting blooms, but its high water needs mean that it won't be appropriate for every landscape. However, if you have a small pond, a kitchen sink set up with a graywater outlet, or even an aquaponics tank, then scarlet monkeyflower can be a viable choice. Scarlet monkeyflower spreads to form dense patches of fuzzy leaves and tubular flowers with a long bloom season. This species is tolerant of both heat and cold; its only limiting factor is water. The smaller seep monkeyflower (*E. guttata*) requires similar growing conditions and is more widespread in the Southwest. The yellow flowers of seep monkeyflower are bee-pollinated, while those of scarlet monkeyflower are frequented by hummingbirds.

Gaillardia pulchella – Asteraceae

firewheel

1′ × 1′

full sun, partial shade

low to medium water needs

flowers year-round

Field guide: Most common on sandy plains and roadsides in Semidesert Grassland and Great Basin Grassland, from 4000 to 6000 feet.

Description: The stunning blooms of this wildflower and its well-deserved popularity in Southwest landscaping have made it ubiquitous in nurseries, even those that seem to studiously avoid carrying useful native plants. With tri-colored maroon, orange, and yellow flowers, firewheel is a worthy addition to any flower bed and may even convert your stubborn neighbors or HOA yard enforcers into native plant enthusiasts. The petals of this plant support caterpillars of the lovely painted schinia moth (*Schinia volupia*), and the foliage is eaten by bordered patch (*Chlosyne lacinia*) butterfly larvae. Though short-lived, firewheel readily reseeds in most landscapes, so it can be expected to be a permanent fixture in your garden, where it can serve as a living advertisement for the benefits of gardening with native plants.

Geranium caespitosum – Geraniaceae

pineywoods geranium (geranio del pino)

1′ × 2′

partial shade

medium water needs

flowers May–September

Field guide: Found in almost all the mountain ranges of the Southwest, especially in Riparian Corridors passing through woodlands, and in Montane Conifer Forest between 5000 and 9000 feet.

Description: Best suited to mid- to high-elevation gardens, this species will do well in the filtered light of trees or on the shady side of a wall. The recurved purple flowers, attractive spreading foliage, and cranesbill seedpods will make this a beloved member of your garden. Geranium flowers are regularly visited by bees and butterflies, so consider including this species in a garden with three-nerve goldenrod (*Solidago velutina*), horsetail milkweed (*Asclepias subverticillata*), and Hooker's evening primrose (*Oenothera elata*) for a range of colors and textures that will provide food for a diverse array of pollinators.

Glandularia bipinnatifida and *G. gooddingii* – Verbenaceae

Dakota mock vervain and Goodding's mock vervain

2′ × 2–3′

full sun, partial shade

low to medium water needs

flowers year-round

Field guide: Mock vervains can be found across almost all the habitats of the Southwest, from 1000 to 10,000 feet.

Description: These two species of mock vervain, found throughout the Southwest, look and behave quite similarly, differing only in Goodding's mock vervain preferring low-elevation landscapes and Dakota mock vervain being found in cooler, high-elevation areas. These plants have one of the longest bloom seasons of any native wildflower and a tolerance for a variety of soils and exposures, which makes mock vervain a necessity for any native plant garden. The ability of mock vervain to reseed in gardens makes up for the fact that this species can be short-lived. The seeds are a food source for a variety of small birds and bees, and butterflies will consistently hang around the pink-to-lavender flowers.

Hedeoma nana – Lamiaceae

dwarf false pennyroyal (orégano chiquito)

1′ × 2′

full sun, partial shade

low water needs

flowers March–October

Field guide: This species can be found in habitats ranging from Arizona Upland Sonoran Desertscrub and Mojave Desertscrub to Montane Conifer Forest. Look for dwarf false pennyroyal on gravelly slopes or in rock crevices from 1000 to 7500 feet.

Description: This member of the mint family is currently underutilized in Southwest landscaping, where its tubular lavender flowers would be a great contributor to increasing backyard hummingbird and butterfly resources. This plant is represented by three subspecies occurring in a broad range of habitats, but all share the same diminutive size and dense growth habit with square stems, opposite leaves, and pleasantly fragrant foliage. Dwarf false pennyroyal is well suited to pollinator gardens, and its short stature means it can be used as a foreground planting backed by taller flowering shrubs.

Heliomeris multiflora – Asteraceae

showy goldeneye (tacote)

3–4′ × 2–3′

full sun, partial shade

low to medium water needs

flowers May–October

Field guide: This species is found in Madrean Evergreen Woodland and Great Basin Conifer Woodland habitats from 4500 feet and into Montane Conifer Forest up to 9000 feet. It's found in most Southwestern mountain ranges on slopes, roadsides, and meadows.

Description: This bushy, multi-branched species can form prolific patches that carpet slopes and meadows in a sea of gold during the monsoon season. Showy goldeneye will thrive in upland and high-elevation landscapes, where it can be used as a border planting, mixed into a meadow garden, or used to restore disturbed habitats. As with many members of the sunflower family, showy goldeneye appeals to numerous butterfly and bee species, and various backyard birds will eat the seeds (and deposit them around your garden in nice piles of nutrients).

Hibiscus coulteri – Malvaceae

desert rose mallow (tulipán)

3′ × 1′

full sun, partial shade

low water needs

flowers year-round following rains (especially March–October)

Field guide: Found in Arizona Upland Sonoran Desertscrub and Semidesert Grassland from 1000 to 4000 feet, where it grows on rocky slopes and sandy wash edges.

Description: Lovers of exotic plants may be surprised to learn that the Southwest is home to several species of native hibiscus. Ours are less gaudy than the cultivars of tropical species commonly found in nurseries, but they provide gorgeous flowers and immense wildlife value while being well-adapted to desert heat. *Hibiscus coulteri* is one of these, with upright, lanky stems that often grow through the branches of trees and shrubs. Following rains, this plant reveals a tulip-like assemblage of overlapping lemon-yellow petals with splashes of crimson near their bases. All our native hibiscus species are pollinated by bees and butterflies and are a food source for several moth and butterfly caterpillars, including Geometer moths (*Geometridae* spp.) as well as mallow scrub hairstreak (*Strymon istapa*) and Arizona powdered skipper (*Systasea zampa*) butterflies.

Hymenoxys hoopesii – Asteraceae

owl's claws

2–3′ × 1–2′

full sun, partial shade

medium water needs

flowers July–September

Field guide: Grows from 7000 to 11000 feet, near streams and moist meadows in Montane Conifer Forest and Subalpine Conifer Forest habitats.

Description: This species is ideal for high-elevation gardens with water-harvesting infrastructure like basins, swales, and gray-water outlets. As large-flowered asters, owl's claws are pollinator powerhouses that are particularly favored by bumblebees (*Bombus* spp.) and butterflies, but are also visited by moths. This species tends to do best in areas where it receives morning sun and afternoon shade or filtered light from an overhanging tree. Pair with other cold-tolerant plants that require extra moisture, such as Rocky Mountain iris (*Iris missouriensis*) and deergrass (*Muhlenbergia rigens*).

Ipomopsis aggregata – Polemoniaceae

scarlet gilia

3′ × 1′

full sun, partial shade

medium water needs

flowers May–September

Field guide: Grows on cliff faces, roadsides, and clearings in Great Basin Conifer Woodland and Montane Conifer Forest from 5000 to 9000 feet.

Description: One of the most brilliant wildflowers in the Southwest mountains, this plant thrives in dry or wet soil and will happily tolerate full sun or light shade. It is short-lived, typically lasting two years, but produces an abundance of seed, and volunteers are likely to follow blooms where conditions are favorable. For the first part of its life, it grows as a rosette of ferny leaves that slightly resemble yarrow. When ready to flower, the plant begins to elongate, and a tall, branched stalk reaching two to three feet will develop and be adorned with tubular blossoms that flare out at the tips like starfish. Typically, the flowers are fiery scarlet with throats mottled red and white, but sometimes individuals or whole populations have white blossoms. Hummingbirds and large butterflies like the two-tailed swallowtail (*Papilio multicaudata*) are frequent visitors. White-lined sphinx moths (*Hyles lineata*) are also attracted to these plants, especially where the white flower variation is present.

Iris missouriensis – Iridaceae

Rocky Mountain iris

1–2′ × 1–2′

full sun, partial shade

high water needs

flowers April–August

Field guide: Grows throughout the Rocky Mountains in boggy meadows and moist stream edges and reaches its southern limit in northern Mexico. Look for it between 6000 and 9500 feet in Great Basin Conifer Woodland and Montane Conifer Forest.

Description: You don't have to be a native plant enthusiast to become enamored with Rocky Mountain iris; the gracefully spreading petals and neat form would make this species as much at home in a formal English manor garden as a pollinator-friendly Southwest landscape. This species requires regular watering, especially from spring to summer while the flowers are developing. This species can be propagated by seed or by digging it up and dividing the bulbs, which is best done after flowering in early fall. Create a mass planting to line a pathway or fill in a terraced bed; or, mix this species in with yellow and red wildflowers for a blend of different foliage textures and bloom colors.

Jatropha macrorhiza – Euphorbiaceae

ragged nettlespurge (jirawilla)

1.5′ × 1.5′

full sun, partial shade

low water needs

flowers May–October

Field guide: Found on gravelly slopes and rock outcroppings in Semidesert Grassland and Madrean Evergreen Woodland from 3500 to 5500 feet.

Description: An underappreciated botanical treasure of the US–Mexico Borderlands, this species deserves wider use in landscaping, where its unusual aesthetic will contribute a distinctive structure to grassland, succulent, and pollinator gardens. Dormant through winter and spring, ragged nettlespurge emerges before monsoon rains, growing from a tuberous, fleshy root to unfurl batwing-like leaves that start out a glossy burgundy before turning rich green. As rains arrive, pink five-petaled flowers appear, offering nectar to the butterflies and bees that seem to appear out of thin air when the humidity spikes. Caterpillars of the cincta silk moth (*Rothschildia cincta*), a beautifully adorned and quite large species, feed on the foliage.

Justicia longii – Acanthaceae

longflower tubetongue

1′ × 1′

full sun, partial shade

low water needs

flowers April–October

Field guide: Look for this plant between 2500 and 4000 feet in the Sonoran Desert of southern Arizona and northern Sonora, Mexico, where it thrives on rocky bajadas and the loose sand of washes.

Description: This can be a difficult species to see because of its diminutive size and night-blooming flowers, which wither and fall as the heat of the morning rises. Ideal for gardeners hoping to attract moths, longflower tubetongue delivers a lot of flowers with not much fuss and will happily grow with more showy species like Gooding's mock vervain (*Glandularia gooddingii*) and fairy duster (*Calliandra eriophylla*) as well as charismatic cacti like fishhook barrel (*Ferocactus wislizeni*) and saguaro (*Carnegiea gigantea*). You may fail to notice longflower tubetongue through most of the year, but when you walk out your door one morning and are greeted by hawk moths (*Sphingidae* spp.) darting between white flowers, you will be grateful to have this species in your garden.

Linum lewisii – Linaceae

blue flax

2′ × 1′

full sun, partial shade

medium to high water needs

flowers April–September

Field guide: This plant is found throughout the American West across a range extending from 3500 to 11,500 feet. It's a common component of mountain meadows, woodland openings, and prairies from Great Basin Grassland to Subalpine Conifer Forest.

Description: Cheery flowers make this a perennial favorite of high-elevation gardeners in the Southwest. The upright form of this species helps it stand out in a prairie planting mixed with grama grasses (*Bouteloua* spp.) and little bluestem (*Schizachyrium scoparium*). This plant can form dense stands that cast a blue haze over grasslands and meadows, a look that can be mimicked in a cultivated landscape. Blue flax has five-petaled flowers followed by roughly globe-shaped seedpods. This versatile plant is an easy and indispensable part of upland Southwestern gardens, where its long bloom season will ensure a steady stream of butterflies and bees visiting your landscape.

Mandevilla brachysiphon – Apocynaceae

Huachuca rock trumpet (hierba de San Juan)

1.5′ × 2′

full sun

low to medium water needs

flowers June–September

Field guide: Found only in the sky islands of southeastern Arizona and southwestern New Mexico, in Semidesert Grassland and Madrean Evergreen Woodland. This species grows between 3500 and 6000 feet on rocky slopes and limestone outcrops.

Description: This species has everything a gardener could want packed into one phenomenal wildflower. If you're looking for drought tolerance, this plant has it, often growing out of bedrock or loose scree and requiring minimal irrigation. If showy blooms are what you're after, then look no further; come summer, Huachuca rock trumpet delights the senses with masses of plumeria-like flowers. They give off a heavenly jasmine aroma that's as likely to draw humans as pollinators. Speaking of pollinators: Butterflies and hummingbirds will visit during the day, and when the sun goes down, sphinx (*Manduca* spp.) and hawk (*Sphingidae* spp.) moths will appear, navigating by the moonlight reflected off the pearly-white blossoms. Huachuca rock trumpet is perfectly suited to well-draining rock gardens and also makes a lovely potted specimen on a patio or by a window, where the fragrance can be appreciated.

Melampodium leucanthum – Asteraceae

blackfoot daisy

6″–1′ × 1–2′

full sun

low water needs

flowers year-round (especially March–November)

Field guide: In habitat, this species is often found on exposed rock, dry slopes, and limestone outcrops in Interior Chaparral, Semidesert Grassland, and Great Basin Grassland communities between 2000 and 6000 feet.

Description: A popular, reliable, and versatile perennial, this species will flourish even in landscapes with poor soils and is ideal for mounded rock gardens. This plant is equally at home in a formal flower garden with mock vervain (*Glandularia* spp.) or a rugged natural planting where it can be interspersed with cacti, yuccas, and agaves. Blackfoot daisy does well in containers, where it will elegantly drape over the edges. A long bloom season and ease of cultivation make it one of the most widely used wildflowers for native gardens. Heavy clays may stunt the growth of this species, and it is worth pruning back periodically to keep a dense form and encourage abundant flowering.

Menodora scabra – Oleaceae

rough menodora

1.5′ × 2′

full sun

low water needs

flowers March–October

Field guide: This widespread species is fond of arid, rocky slopes and can be found from 2000 to 7500 feet, in habitats from Arizona Upland Sonoran Desertscrub to Madrean Evergreen Woodland and Great Basin Conifer Woodland.

Description: This hardy wildflower is recognizable by its yellow five-petaled flowers and its fruit, which consists of paired orbs that split open at the top like Easter eggs to reveal the pitted, teardrop-shaped seeds. It is surprising that such a lovely plant is not more widely cultivated in the horticulture trade. Rough menadora is a dense plant with numerous branches bearing narrow, alternate leaves. It will be right at home in a cactus-centric landscape interspersed with decorative boulders and low shrubs.

Mirabilis longiflora – Nyctaginaceae

long-flowered four o'clock (maravilla)

2–3′ × 2–3′

full sun, partial shade

low water needs

flowers May–October

Field guide: Look for this plant on rocky slopes and sandy canyon edges in the mountains of central and southern Arizona and New Mexico between 3000 and 7500 feet. It is most often found along Riparian Corridors from Semidesert Grassland and Great Basin Grassland habitats and into Montane Conifer Forest.

Description: This captivating monsoon wildflower can be cultivated in a variety of conditions and at a wide range of elevations. It rises from fleshy roots and can grow up to five feet tall depending on moisture availability. The heart-shaped leaves are attractive and serve as an important food source for caterpillars of the white-lined sphinx moth (*Hyles lineata*). But the real stars of the show are the blooms: several-inch long floral tubes enclosed in hairy bracts, which terminate in fragrant white blossoms with decorative purple stamens. The orange-blossom aroma and moonlight-reflectivity of the white flowers attract night-flying insects, in particular hawk moths (*Sphingidae* spp.). After frosts, this plant dies back to its roots and lays dormant until rising temperature and humidity encourage it to send up new growth again.

Mirabilis multiflora – Nyctaginaceae

Colorado four o'clock

6″–1′ × 1–2′

full sun, partial shade

low water needs

flowers April–October

Field guide: Found across much of the Southwest on sandy flats and rocky slopes from 2500 feet in Arizona Upland Sonoran Desertscrub and Mojave Desertscrub to 8500 feet in Montane Conifer Forest.

Description: Gorgeous deep purple flowers and a dense mounding habit make this plant a lovely choice for pollinator gardens. The blooms appear in the evening and fade by late morning, so the primary pollinators are moths, but hummingbirds and butterflies also visit. In winter this species dies back to an underground tuber, but don't fear; warm temperatures will bring new growth and prolific flowering. Because of its deciduous nature, it is wise to surround this plant with evergreen flowers such as blackfoot daisy (*Melampodium leucanthum*) and mock vervain (*Glandularia gooddingii* or *G. bipinnatifida*). Plant in well-draining soil where the tuber won't rot and consider planting on the edge of a low rock wall where the stems can drape languidly over the edges.

Oenothera cespitosa – Onagraceae

tufted evening primrose

6″ × 1′

full sun, partial shade

low water needs

flowers March–September

Field guide: This species can be found on rocky slopes, rolling plains, and mesa tops from 3000 feet in the upper reaches of the Southwest's deserts to around 7500 feet in Montane Conifer Forest.

Description: A welcome addition to most any landscape, this plant has large, four-petaled blooms that resemble satellite dishes. They open late in the day, and the white petals reflect moonlight, which helps guide night-flying pollinators, including the beguiling white-lined sphinx moth (*Hyles lineata*). This wildflower is particularly effective when planted en masse or used to line pathways where it can fill in and form an uninterrupted groundcover of soft, lance-shaped leaves. Intersperse with other moth-pollinated species like sacred datura (*Datura wrightii*) and yuccas (*Yucca* spp.), or plant in clumps in a meadow setting with beardtongues (*Penstemon* spp.) and grasses.

Oenothera elata – Onagraceae

Hooker's evening primrose

3–6′ × 1–3′

full sun, partial shade

medium water needs

flowers year-round (especially June–October)

Field guide: Grows in meadows, stream edges, roadside ditches, and wooded slopes from 3500 to 9000 feet in a variety of habitats.

Description: This primrose starts life as a rosette of flat leaves hugging close to the soil. It begins to mound up at the center gradually, producing one or more upright stems that have thin leaves and an abundance of tightly closed buds and may reach several feet high. As night falls these buds begin to loosen, revealing the petals beneath, before exploding into a profusion of banana-yellow blooms that are a beacon to passing moths. Like other primroses, this species is a larval food for the white-lined sphinx moth (*Hyles lineata*) as well as flower moths in the genus *Schinia*. Plant several Hooker's evening primroses together with sacred datura (*Datura wrightii*) and long-flowered four o'clock (*Mirabilis longiflora*) to enhance the value of your garden to night-flying pollinators.

Penstemon palmeri – Plantaginaceae

Palmer's penstemon

1.5′ × 1′

full sun, partial shade

medium water needs

flowers May–July

Field guide: Grows on rocky slopes and in fields of volcanic stone between 3500 and 7500 feet. Look for it in the Interior Chaparral around Prescott and Sedona, Arizona; in the Great Basin Conifer Woodland near St. George, Utah; and around the Mojave National Preserve in eastern California.

Description: A stunning wildflower with large, bell-shaped blossoms. The petals are a faded pink-bubblegum color, with a golden-haired "tongue" that sticks out as an enticement to passing pollinators. The blooms are carried in densely packed rows on stalks that can rise several feet above the gray-green leaves near the base of the plant. With its large size and highly fragrant blooms, this species is incredibly distinctive in habitat or in a garden. You can underplant Palmer's penstemon with blackfoot daisy (*Melampodium leucanthum*) and place it against a backdrop of large shrubs like sugar sumac (*Rhus ovata*) and oneseed juniper (*Juniperus monosperma*). Although short-lived, this species tends to reseed and establish a population in favorable locations.

Penstemon parryi – Plantaginaceae

Parry's penstemon (pichelitos)

1–2′ × 1′

full sun, partial shade

low water needs

flowers March–May

Field guide: Look for this plant on rocky slopes and sandy washes in Arizona Upland Sonoran Desertscrub, Lower Colorado River Sonoran Desertscrub, Semidesert Grassland, and Madrean Evergreen Woodland habitats from 1500 to 5000 feet.

Description: A spectacular addition to the spring wildflower shows that follow winter rains, this species thrives in the low deserts of the Southwest. From a rosette of pale foliage, one or more tall stalks emerge and give rise to tubular pink flowers that are sure to attract hummingbirds and other pollinators, including moths. Though individual plants are short-lived, they are reliable reseeders that become more and more abundant over time. A mixture of this plant with brittlebush (*Encelia farinosa*) and apricot globemallow (*Sphaeralcea ambigua*) works well to add texture and immediate color to landscapes while newly planted shrubs and trees are still growing. This species adds an early spring pop of color to succulent gardens before the flowering of cacti and agaves. Parry's penstemon is like a nice black shirt: It goes with everything.

Pericome caudata – Asteraceae

mountain tail leaf (yerba del chivato)

3–4′ × 3–4′

full sun, partial shade

low to medium water needs

flowers July–October

Field guide: This species is common on rocky slopes, in meadows, and along sandy roadsides in Great Basin Conifer Woodland and Montane Conifer Forest between 6000 and 9000 feet.

Description: This plant's copious golden flowers appear as other shrubs go dormant in fall, making it an important late-season nectar source. While lacking the alluring ray petals that are characteristic of other members of the sunflower family, this species makes up for it with sheer numbers of blossoms. It can easily reach three to four feet in height and is recognizable even before flowering due to its distinctive triangular leaves that come to a long narrow point. Plant mountain tail leaf in a rock garden between shorter flowers like blue flax (*Linum lewisii*) and dwarf false pennyroyal (*Hedeoma nana*) and in front of larger shrubs and trees like hairy mountain mahogany (*Cercocarpus breviflorus*) and ponderosa pine (*Pinus ponderosa*).

Psilostrophe cooperi – Asteraceae

paper flower

1–2′ × 1–2′

full sun

low water needs

flowers March–October

Field guide: This species frequents sandy wash edges and gravelly flats and slopes in Arizona Upland Sonoran Desertscrub and Mojave Desertscrub and into Semidesert Grassland from 2000 to 5000 feet.

Description: Tolerant of full sun or light shade, this is a hardy and versatile wildflower with an unfussy nature, making it a desirable addition to desert gardens. The foliage has a gray-green cast and tends to form a low mound of spreading branches topped by yellow flowers with petals bent backward. The blooms provide food for bees and butterflies and seem to hold a particular appeal for the brilliant, emerald-colored sweat bees (*Melissodes* spp.) that frequent desert gardens. Interplant with other low-mounding species like Goodding's mock vervain (*Glandularia gooddingii*) and desert marigold (*Baileya multiradiata*) in a wildflower or succulent garden.

Rudbeckia laciniata – Asteraceae

green-head coneflower

2–3′ × 1–2′

full sun, partial shade

medium to high water needs

flowers June–October

Field guide: Prefers moist habitats and is often found in Riparian Corridors along streams or in meadows from 5000 to 9000 feet in Montane Conifer Forest.

Description: This superb pollinator attractor has large composite flowers that act as a perfect landing surface for passing butterflies. The plant spreads by underground runners, and the flowers top upright stems with lobed leaves that are relished by caterpillars of the silvery checkerspot butterfly (*Chlosyne nycteis*) and wavy-lined emerald moth (*Synchlora aerata*), among others. Consider planting in a basin or near a gray-water outlet where supplemental irrigation can easily be provided in summer months. The erect form of this wildflower looks great in combination with straight, upright trees like Ponderosa pine (*Pinus ponderosa*), Arizona cypress (*Hesperocyparis arizonica*), and velvet ash (*Fraxinus velutina*).

Ruellia ciliatiflora (syn. *Ruellia nudiflora*) – Acanthaceae

violet wild petunia (rama de toro)

1′ × 1′

full sun, partial shade

medium water needs

flowers March–November

Field guide: Usually found along sandy washes and ciénaga edges under a thick mesquite canopy or near wetlands in Arizona Upland Sonoran Desertscrub and Semidesert Grassland between 2000 and 4500 feet.

Description: One of our very best wildflowers for shade, especially in low-elevation gardens where most species are better adapted to sunny conditions. Quick to reseed and establish a self-sustaining population, this species has purple, bell-shaped blossoms that attract myriad pollinators, including bumblebees (*Bombus* spp.), queen butterflies (*Danaus gilippus*), and Anna's hummingbirds (*Calypte anna*), among others. Planted in the deep shade of a mature mesquite or on the north side of a house, it will need a little extra water in summer to ensure ample blooms through the warm months. By combining violet wild petunia with other shade-amenable species like leadwort (*Plumbago zeylanica*), vine mesquite grass (*Hopia obtusa*), and longflower tubetongue (*Justicia longii*), you can make a gorgeous and ecologically functional planting in an otherwise difficult shady location.

Salvia arizonica – Lamiaceae

Arizona sage

1–2′ × 2–3′

partial shade, shade

medium water needs

flowers May–October

Field guide: Generally found between 7000 and 9500 feet as an understory plant in Montane Conifer Forest in the sky island mountain ranges of southeastern Arizona.

Description: This shade-tolerant plant can be used to fill in beds around mature trees in high-elevation gardens. Lack of light won't keep it from producing the violet or lavender flowers that make it such an attractive part of a landscape. While this species is not well suited to desert gardens, it is ideal for cold areas and will easily withstand frost and snow. Identify Arizona sage when not in flower by its small stature, square stems, and opposite, triangular leaves with a light sage fragrance. Combine with other montane species like mountain tail-leaf (*Pericome caudata*), tasselflower brickellbush (*Brickellia grandiflora*), and mountain muhly (*Muhlenbergia montana*), and place a ponderosa pine (*Pinus ponderosa*) or bigtooth maple (*Acer grandidentatum*) nearby to provide shade.

Senna covesii – Fabaceae

desert senna (ejotillo)

1–2′ × 1–2′

full sun

low water needs

flowers year-round (especially March–October)

Field guide: Found on sandy flats and gravelly slopes in Arizona Upland Sonoran Desertscrub, Lower Colorado River Sonoran Desertscrub, and Mojave Desertscrub into Semidesert Grassland between 1000 and 5000 feet.

Description: A common desert wildflower that can be abundant in years with above-average rainfall, this species is prized because it is not restricted to flowering in spring or summer but will bloom essentially all year when conditions of temperature and humidity allow. Identify this species by its short stature, fuzzy hairs on the leaves, three to four pairs of leaflets, bright yellow five-petaled flowers, and long seedpods that rattle and shake as you brush them. Easily planted from seeds or starts, desert senna will have no problem in full sun and can be grown on clay or sandy soils (though it prefers the former). A particularly effective landscape element when combined with other wildflowers like desert marigold (*Baileya multiradiata*), apricot globemallow (*Sphaeralcea ambigua*) and Gooding's mock vervain (*Glandularia gooddingii*).

Silene laciniata – Caryophyllaceae

cardinal catchfly (clavel de monte)

2–4′ × 1–3′

full sun, partial shade

medium water needs

flowers April–October

Field guide: Look for this plant on rocky hills, roadsides, and openings from 5500 to 10,000 feet in Madrean Evergreen Woodland, Great Basin Conifer Woodland, and Montane Conifer Forest. This is an understory species in eastern Arizona and western New Mexico.

Description: A strikingly odd wildflower with astounding scarlet blooms and sticky foliage and stems, cardinal catchfly makes an interesting specimen for high-elevation gardens. In the Southwest, this species does best with some late-afternoon shade. The blooms attract hummingbirds and the adhesive stems can trap unwary insects—possibly an adaptation to prevent these same insects from dining on the plant. The unusual and distinctive flowers of this species immediately draw the eye, and the plant can be interspersed among other wildflowers in a north- or east-facing rock garden or terrace.

Solidago velutina – Asteraceae

three-nerve goldenrod

2–3′ × 2–3′

full sun, partial shade

medium to high water needs

flowers June–October

Field guide: Found from 3500 to 8500 feet in grassy meadows, roadside ditches, streamsides, and slopes in wooded Riparian Corridors, Semidesert Grassland, Madrean Evergreen Woodland, Great Basin Conifer Woodland, and Montane Conifer Forest habitats.

Description: The Southwest is lucky to have several species of goldenrods, which are among the most highly touted pollinator plants globally, and this is one of the most common. Despite its preference for wet places, this versatile, relatively fuss-free plant is not overly picky about habitat. It forms patches that spread out from a creeping root system with thin stems and a fine fuzz of hair on alternating lance-shaped leaves. The profuse golden flowers are all borne on one side of a nodding inflorescence that is unmistakable against the mix of grasses and subshrubs with which goldenrod is often associated. One of our best fall nectar sources, it adds a much-needed pop to a landscape with its mass of late blooms just as other species are fading.

Sphaeralcea ambigua – Malvaceae

apricot globemallow (mal de ojo)

3′ × 3′

full sun, partial shade

low water needs

flowers year-round (especially spring)

Field guide: Found on rocky slopes and sandy wash edges from just above sea level to 3500 feet, in Arizona Upland Sonoran Desertscrub, Lower Colorado River Sonoran Desertscrub, and Mojave Desertscrub into Semidesert Grassland.

Description: Common in nearly every part of the Southwest, globemallows include many highly variable species that occupy niches from low-desert flats to cool mountain meadows. In Arizona deserts, this genus is most prominently represented by apricot globemallow. The flowers vary from orange to pink, lavender, red, or white. All globemallows share the basic form of five-petaled cups borne singly or in clusters on upright stems coated with fine hairs. These flowers are primarily pollinated by specialized chimney bees (*Diadasia* spp.) but also attract other bees as well as butterflies; the leaves are a larval food source for checkered skippers (*Pyrgus* spp.) and painted lady (*Vanessa cardui*) butterflies. The seeds attract a variety of backyard birds, and desert tortoises happily munch on the blooms.

Thelesperma megapotamicum – Asteraceae

Hopi tea greenthread (cota)

1–2′ × 3–4′

full sun, partial shade

low to medium water needs

flowers May–October

Field guide: Found in eastern Arizona, most of New Mexico, and into southern Utah and Colorado, with a broad geographic and topographic distribution. Often associated with open plains, mesa tops, and disturbed habitats from 3500 to 9500 feet, from Chihuahuan Desertscrub and Great Basin Desertscrub up to Montane Conifer Forest.

Description: Suitable for most Southwestern gardens, this shrub consists of wiry stems and thread-like leaves, which grow into sprawling mounds terminating in clusters of showy golden blossoms that attract pollinators, including butterflies and bees. If you have heard of this species, it is most likely due to its renown as a tasty medicinal tea used widely by people in the Southwest for millennia and increasingly common at stores and farmer's markets. Try planting it alongside a drought-tolerant evergreen like oneseed juniper (*Juniperus monosperma*) and a wildlife-friendly shrub such as three-leaf sumac (*Rhus aromatica* var. *trilobata*) to create a beautiful and ecologically functional plant guild.

Thymophylla pentachaeta – Asteraceae

golden fleece (parralena)

6″ × 1′

full sun, partial shade

low water needs

flowers year-round (especially March–October)

Field guide: Found across every state in the region, this widespread Southwestern wildflower grows on gravelly flats and limestone slopes in desertscrub and grassland habitats from 2000 to 5500 feet.

Description: This species is a gift that keeps on giving, with prolific seed production that ensures not only volunteers in your garden but in your neighbor's garden too. The little clumps of green foliage and masses of yellow flowers carpet the ground and attract pollinators and birds, who help distribute the seeds. The tangy, earthy-smelling leaves are also a larval food source for the dainty sulphur butterfly (*Nathalis iole*). Extremely easy to grow, it can be seeded or planted into a wildflower garden or used to fill space between cacti and succulents. Four varieties are found in the Southwest, but the differences between them are probably too fine to be of interest to most gardeners; however, purists may want to research and seek out their local variety.

Tetraneuris acaulis – Asteraceae

angelita daisy

6″–1′ × 9″–1′

full sun, partial shade

low water needs

flowers April–July

Field guide: Mostly absent from the southern parts of Arizona and New Mexico, this species becomes common farther north toward the mesas, plains, and woodlands of the upper Southwest from 4000 to 8500 feet.

Description: Angelita daisy is one of the most common native wildflowers in the Southwest's horticulture trade. Its popularity derives from its ease of cultivation and long-blooming yellow flowers. Very cold-tolerant, this is a reliable choice for mid- to high-elevation gardens where it can be planted at the bases of larger shrubs or mixed into a wildflower and grass bed with other short species so that it isn't lost in the mix. Good companion plants might include pineywoods geranium (*Geranium caespitosum*), tufted evening primrose (*Oenothera cespitosa*), and rice-grass (*Achnatherum hymenoides*).

Zinnia grandiflora – Asteraceae

prairie zinnia

8″ × 1′

full sun

low water needs

flowers April–October

Field guide: Common across much of the Southwest between 3000 and 7500 feet on exposed slopes, alkaline flats, or limestone outcrops. Usually found in Semidesert Grassland, Great Basin Grassland, and Interior Chaparral in Arizona, Colorado, New Mexico, and west Texas.

Description: This vibrant groundcover has lovely flowers and a low-mounding habit that makes it ideal for edging a pathway or planting at the base of a low wall. Accentuate its qualities by combining it with other low-growing wild-flowers like tufted evening primrose (*Oenothera cespitosa*) and blackfoot daisy (*Melampodium leucanthum*). It's a valuable nectar source for a variety of bees and butterflies, and its golden flowers with orange centers pop against a backdrop of light granite or red sandstone. Planted at the edge of a rock garden or in a pot, prairie zinnia will cascade gracefully in a spray of wiry green foliage and brilliant blossoms. This species can cover large swaths of ground, a habit it will hopefully mimic in your garden. At lower ele-vations this species is replaced by hardy, white-flowered desert zinnia (*Zinnia acerosa*).

Cacti and Succulents

One of the great joys of landscaping in the Southwest is the distinctive architectural forms that our native cacti and succulents bring to a garden. Cacti can seem difficult to work with for new gardeners, but they are essential to Southwestern landscapes and you will learn to love them, spines and all. These plants range from tiny pincushions barely peeking above the ground to succulent trees that dominate a landscape. Though most people associate cacti and succulents with the low desert, the Southwest furnishes a range of species appropriate for almost any garden, able to tolerate sweltering heat or frigid cold in a variety of exposures. Cacti and succulents offer year-round interest due to their bizarre and sometimes grotesque forms, along with stunning flowers that rival those of any shrub or perennial. These plants are often slow-growing, so it is helpful to do some research to assess the ultimate size of the succulent you are looking at, ensuring that you aren't forced to move an unwieldy, spine-covered cactus or agave in the future. Succulents are uniquely well-adapted to the arid climate of the Southwest, with shallow root systems that capture water as soon as it hits the ground and the ability to store moisture internally, which allows them to survive long periods of drought. The ability to store moisture does mean that cacti and succulents can be rot-prone, meaning that they require coarse, well-draining soil to thrive, and they will tend to languish or die in heavy clays. Planting a gaudily-spined cactus, a showy ocotillo, or an intriguing agave will put a distinctly Southwestern spin on your landscape that requires minimum maintenance to thrive.

Agave deserti – Asparagaceae

desert agave (amul)

2.5′ × 3′

full sun

low water needs

flowers May–July

Field guide: Grows in harsh exposures on nutrient-poor soils from 1000 to 3000 feet, on rocky hillsides and sandy wash edges in Lower Colorado River and Arizona Upland Sonoran Desertscrub. If you see a sizable agave with numerous offsets in the lower Sonoran Desert, it is most likely this species.

Description: Distinguishable from other Southwestern agave species by its habit of readily producing many-headed clusters; its waxy, ashen leaves; curved marginal teeth; and distinctive marks on the bottom of each leaf. Like most agaves, this species is monocarpic, meaning it will live for several years, bloom once, and then die in a blaze of golden-flowered glory set to the symphony of insects, bats, and birds buzzing, chirping, and humming around the blossoms. As one rosette withers, others in the cluster continue to grow. Although this species readily endures full sun and high heat, well-draining soil is a must, and occasional additional water in summer supports robust growth.

Agave palmeri – Asparagaceae

Palmer's agave (lechuguilla)

4′ × 4′

full sun, partial shade

low water needs

flowers June–August

Field guide: Typically found on gravelly slopes and rock outcrops in Semidesert Grassland and Madrean Evergreen Woodland between 3000 and 6000 feet, this sky-island species may grow out in the open among grasses or directly on cliff faces in dense woodland where there is more competition for light.

Description: Variable in size but usually a large rosette of stiff, guttered leaves that direct water to the root system, this species may live for decades before finally sending a flower stalk ten to fifteen feet in the air with clusters of waxy, green-pink flowers filled with nectar. The flowers are important to nectar-feeding bats; farther north, the primary pollinator is hummingbirds. Elegant and imposing, it has conspicuous bud prints on its blue-green leaves, each of which is lined with dark recurved teeth and tipped with a stiff spear-like point. At lower elevations, this agave benefits from light shade. You can plant it under velvet mesquite (*Neltuma velutina*) or on the east side of a structure, where it will receive afternoon shade. In mid-elevation gardens, plant it in full sun mixed with wildflowers and perennial grasses. Needs ample room to grow so that walking or working near the plant won't present a safety hazard (due to its size and armament).

Agave parryi – Asparagaceae

Parry's agave (mescal)

2′ × 4′+ (spreads by roots)

full sun, partial shade

low water needs

flowers June–August

Field guide: Occupies dry, rocky hillsides and cliff faces in Semidesert Grassland, Madrean Evergreen Woodland, Interior Chaparral, and Montane Conifer Forest between 4500 and 7000 feet. Visitors to our region are often surprised to see woodland species such as pines and oaks growing alongside cacti and succulents.

Description: Grows in many forms, from solitary rosettes on rock outcrops to mass clusters of hundreds of clones covering grassy slopes. Generally, these are stout agaves, with tightly packed leaves that make them resemble spiny artichokes. Thick stalks rise in summer to culminate in a dazzling display of golden flowers that instigate a riot of pollinators eager to reach them. Perfect for planting on slopes where their spreading root system will help stabilize the soil, they also work well in containers, especially shallow bowl-style pots, which they fill with numerous offsets. Wherever you plant this species, ensure the soil has good drainage to avoid rot. The variety that occurs in southeastern Arizona (var. *huachucensis*) is caterpillar food for the Huachuca giant skipper butterfly (*Agathymus evansi*). Other varieties may also host different butterflies and moths.

Agave utahensis – Asparagaceae

Utah agave

1′ × 1′

full sun, partial shade

low water needs

flowers June–August

Field guide: Grows in hard-to-reach gravel slopes, boulder fields, mesas, and cliffs in Mojave Desertscrub and Great Basin Desertscrub and into Great Basin Conifer Woodland. Found in out-of-the-way places between 3000 and 7500 feet, so seeing it in habitat often involves a long hike and steep scramble; this has preserved some populations from the unwanted attention of plant poachers.

Description: In the mountain ranges ringing the Mojave Desert, this species clings to cliffs, pokes out from loose gravel, and thrives on limestone outcrops. Its small size means that gardeners should plant it where it can be easily seen: in a raised rock or crevice garden or on a mound in well-draining soil. The small green-yellow flowers line a tall, thin flower stalk where they are prominently displayed to pollinators. The leathery foliage of this species is the sole food plant for caterpillars of the Mojave giant skipper butterfly (*Agathymus alliae*). This cold-tolerant agave is perfect for upland and high-elevation gardens. Use caution and ask questions about provenance when purchasing this plant, as poachers have targeted this species to feed the succulent-collecting frenzy fueled by social media and disreputable nurseries.

Carnegiea gigantea – Cactaceae

saguaro (sahuaro)

30–40′ × 10–15′

full sun

low water needs

flowers April–June

Field guide: Typically the tallest thing growing on the rocky slopes, gravelly flats, and rock outcrops of Lower Colorado River Sonoran Desertscrub and Arizona Upland Sonoran Desertscrub, this species is easy to find. Some of the densest saguaro forests anywhere on the planet exist just outside Tucson, Arizona, but any drive along I-8 or I-10 will yield ample opportunity to marvel at this species' distinctive shapes and massive size. Saguaros prefer warm locations between 500 and 3500 feet where frost is rare.

Description: No plant species is more iconic to the Sonoran Desert! These graceful giants prove that it's possible to thrive in unforgiving environments. Saguaros are very slow-growing, eking out only a few centimeters of growth per year, but decent-sized plants are usually available in nurseries. If you plan to buy a salvaged or transplanted individual, confirm that it has an official transit tag so you don't purchase an illegally transported specimen. Young saguaros typically grow under the canopy of a "nurse tree" that helps protect them from extremes of heat and cold. Foothill paloverde (*Parkinsonia microphylla*), velvet mesquite (*Neltuma velutina*), or ironwood (*Olneya tesota*) will work well for this purpose. Plant with a long-term vision in mind; these plants become massive with age, and your grandkids may not want to move a two-ton cactus. The white flowers appear in late spring and attract a variety of pollinators, as do the tasty fruits, which are full of tasty maroon pulp and nutritious seeds. Gilded flickers (*Colaptes chrysoides*) and Gila woodpeckers (*Melanerpes uropygialis*) carve nest cavities into saguaros that will subsequently be used by other bird species over the years, turning a mature saguaro into a desert apartment building. With a lifespan of up to 175 years, a saguaro is the ultimate long-term investment in your landscape, and no Sonoran Desert garden should be without one.

Cylindropuntia acanthocarpa – Cactaceae

buckhorn cholla (choya)

4–6′ × 4–6′

full sun

low water needs

flowers March–June

Field guide: Inhabits valley floors and rocky hillsides in the Sonoran and Mojave deserts into Interior Chaparral and Semidesert Grassland, between 500 and 5000 feet on low hills or in sandy flats.

Description: Understanding the numerous cholla species and their multitudinous hybrids is a task that maddens even seasoned botanists. However, buckhorn cholla genuinely stands out. This species grows across much of western and central Arizona and into southern Nevada and Utah. It is distinctive for its shrubby form, the purple color of its branches when stressed, and the dense, spiny covering on the flower buds and fruits. The flower color can be variable, ranging from saffron-yellow to deep wine-red with shades of orange and pink between. Though not fussy about soil, this species benefits from good drainage and will thrive in a cactus garden with fishhook barrel cactus (*Ferocactus wislizeni*), ocotillo (*Fouquieria splendens*), and Engelmann pricklypear (*Opuntia engelmannii*), where its trellis-like branches add a striking architectural element. The spiny branches are a food source for the staghorn cholla moth (*Euscirrhopterus cosyra*), and you can sometimes find branch tips that have been eaten by this species and resemble intricate skeletons.

Cylindropuntia arbuscula – Cactaceae

Arizona pencil cholla (siviri)

3–6′ × 3–6′

full sun

low water needs

flowers April–June

Field guide: Forms dense patches between 1000 and 3500 feet in the sandy washes and gravelly flats of Arizona Upland Sonoran Desertscrub and is often encountered where rocky slopes meet dry washes; you can also see it in flat plains with creosote (*Larrea tridentata*) and saguaro (*Carnegiea gigantea*).

Description: This species forms a Medusa's head of thin, cascading branches armed with downward-pointing, flaxen-yellow spines. This plant has two basic forms: upright and tree-like or low and bushy. Most specimens will be grown from cuttings, and knowing the form of the parent plant will help you design around this hardy and beautiful cholla. The spring-blooming flowers are typically neon yellow with flecks of red but occasionally are solid scarlet. Because of its dense growth form, Arizona pencil cholla provides good cover for birds, particularly Gambel's quail (*Callipepla gambelii*). The smaller Christmas cholla (*Cylindropuntia leptocaulis*) is differentiated by its thin, twig-like branches and red fruits that persist through winter. This species tends to shed branches over time, so plant somewhere without much traffic to avoid the risk of these ending up in someone's foot.

Cylindropuntia fulgida – Cactaceae

chain-fruit cholla (choya)

4–12′ × 4–12′

full sun

low water needs

flowers April–September

Field guide: Look for this species in sandy flats and gravelly slopes in Arizona Upland Sonoran Desertscrub, Lower Colorado River Sonoran Desertscrub, and Semidesert Grassland between 500 and 4000 feet. It often forms massive colonies (mostly clones propagated by stems rooting into the ground) that are easy to find but tough to get through.

Description: Along with the similar but smaller teddy bear cholla (*Cylindropuntia bigelovii*), this is perhaps the most notorious of all Southwestern cacti. Often referred to as "jumping cholla," this plant's branches detach at the slightest touch, and extracting the barbed spines and sharp hairs known as *glochids* from one's extremities is an unforgettable experience. Though they are much maligned and require cautious handling, chain-fruit chollas make fantastic landscape plants when placed well away from heavy-traffic areas. Not only do the vibrant pink flowers attract bees, but the heavily armed branches are a favorite of cactus wrens (*Campylorhynchus brunneicapillus*), who construct their funnel-shaped nests in this place where few predators will follow. Employ chain-fruit cholla as part of a wildlife hedge or security planting, but be sure to use tongs and great care when handling this species.

Cylindropuntia imbricata – Cactaceae

tree cholla (tasajo)

3–12′ × 3–9′

full sun, partial shade

low water needs

flowers May–August

Field guide: This plant's wide geographic range spans Chihuahuan Desertscrub and Great Basin Desertscrub into Great Basin Grassland, Madrean Evergreen Woodland, and Great Basin Conifer Woodland from 2500 to 7000 feet. Look for these upright cacti on gravelly slopes and well-draining flats.

Description: One of the most common and largest chollas in the Southwest, this species' imposing stature and gorgeous magenta flowers warrant a spot in the yard of any cactus enthusiast. Upright and sometimes reaching the height of a small tree, it can be planted against a wall to provide structure and contrast or used as the centerpiece of a cactus garden. In extreme cold or drought, the branches will often turn from green to purple and droop like a spine-covered willow. The bee-pollinated blooms can appear throughout summer and are followed by plump yellow fruits that persist on the plant, lending the branches a pop of color long after the flowers have withered. Plants at the western edge of this species' range are known as cane cholla (var. *spinosior*) and are similar in having a single trunk but are generally smaller than plants farther north and east.

Cylindropuntia whipplei – Cactaceae

Whipple's cholla

2–4′ × 2–5′

full sun

low water needs

flowers May–July

Field guide: Occurs in sandy plains and gentle slopes in Interior Chaparral, Great Basin Desertscrub, Great Basin Grassland, and Great Basin Conifer Woodland between 4500 and 7000 feet. You can see good examples around the Grand Canyon in Arizona and the Desert National Wildlife Refuge in Nevada.

Description: This plant can form low mats that hide beneath grasses and shrubs, waiting to snag unaware passersby, or it can grow into a low shrub densely armed with silvery spines that have warty yellow fruits persisting on the branch tips. It's one of our most frost-tolerant cacti and will be appropriate for almost any high-elevation garden. The fluorescent yellow flowers cover the plant en masse in spring and early summer, often humming with the activity of bees who can be seen covered in pollen, frolicking inside the blooms. Try planting this in a raised planter where the branches cascade over the edges, or mix into a garden with banana yucca (*Yucca baccata*) and pinyon pine (*Pinus edulis*). Like many other chollas, this species has extrafloral nectaries that provide a sugary reward to ants who patrol the plant and drive away would-be herbivores.

Dasylirion wheeleri – Asparagaceae

desert spoon (sotol)

5–6′ × 5–6′

full sun, partial shade

low water needs

flowers May–July

Field guide: This easy-to-identify succulent inhabits rocky slopes in Chihuahuan Desertscrub, Semidesert Grassland, and Madrean Evergreen Woodland. In the foothills of mountain ranges in these habitats, you may see hundreds of desert spoons silhouetted against the skyline, flower stalks pointing every which way. Prefers elevations between 3000 and 6500 feet in habitat, so is quite cold-tolerant, but will also take low-desert heat.

Description: A distinctive succulent that's very popular in Southwestern gardens but is sometimes planted without regard for its full size, desert spoon is often subject to abusive pruning practices. This species is reminiscent of a trunkless yucca and consists of many thin, strap-like blue leaves lined with razor-sharp teeth on the margins. Eventually, the plant takes on a globe shape and (after a few years) will begin to send up tall, asparagus-like flower stalks with pollen- and seed-bearing blooms borne on separate plants. This versatile landscaping plant is equally happy in low deserts and cool mountain gardens. It can be grown as a centerpiece in a wildflower bed or a decorative element in a cactus garden. The best tip for this species? Don't prune the leaves off! If pruned like a palm, with just a handful of wispy strands left at the top of a barren trunk, it will be exposed to sun and frost damage, and its elegant aesthetic will be destroyed.

Echinocereus coccineus – Cactaceae

claret cup hedgehog (espinas escarlata)

1–1.5′ × 1–3′

full sun, partial shade

low water needs

flowers April–June

Field guide: Grows in rocky soils and outcrops in every Southwestern vegetation type, though it is less common in Desertscrub communities. Found from around 3500 to 9000 feet, this adaptable species is appropriate for almost any garden in the region. These plants often grow on bedrock and are a regular feature of cliffsides around canyons.

Description: The most common member of a large and complex group of hedgehog cacti, all of which have red, cup-shaped flowers and densely clustering forms. The minute differences between species are of little importance to most gardeners, but given the diverse bioregions and habitats where these plants occur, be sure to source specimens grown from seed collected as near to your area as possible. This species' diverse forms range from stout clumps with dense spines half-buried in fallen leaves to majestic, dome-shaped clusters of dozens of branches with widely spaced spines sprouting from near-vertical cliff faces. However, almost all have gorgeous scarlet flowers that are adapted for hummingbird pollination. Most hummingbird flowers are narrow and tubular, but these flowers are sized for the whole heads of hummingbirds, who will eagerly brave the sharp spines to plunder the nectar-rich flowers. The distinctive clumping form and stop-in-your-tracks blooms of claret cup hedgehog will be a welcome addition to your landscape.

Echinocereus fasciculatus – Cactaceae

pinkflower hedgehog cactus (pitahayita)

6″–1′ × 1–2′

full sun, partial shade

low water needs

flowers March–June

Field guide: Likes rocky slopes, gravel flats, and crevices in canyon walls in Arizona Upland Sonoran Desert, Semidesert Grassland, and Interior Chaparral habitats between 2500 and 5000 feet. Because of its low stature and unassuming aesthetic, you do not give this plant more than a passing glance. But when in bloom, its satellite dish–shaped flowers make it both easy and extremely rewarding to find.

Description: Stout and covered in a jacket of sharp spines, pinkflower hedgehog might escape your notice until the namesake flowers explode into glorious bloom in spring. This plant follows its blooms with plump red fruits filled with a tasty white pith similar to the inside of a dragonfruit (both are cacti). If planted in well-draining soil, pinkflower hedgehog can develop into a clump of up to thirty branches, all tipped by show-stopping flowers that almost entirely obscure the top of the plant.

Echinocereus fendleri – Cactaceae

Fendler's hedgehog cactus

6″–1′ × 6″–1′

full sun, partial shade

low water needs

flowers April–June

Field guide: Found at elevations of 3000 to 8000 feet on sandy flats and gravelly slopes in Arizona Upland Sonoran Desertscrub, Chihuahuan Desertscrub, Great Basin Desertscrub, and Semidesert Grassland into Great Basin Conifer Woodland and Montane Conifer Forest. You may find this plant tucked under the base of a tree or poking out of a sheer cliff face.

Description: This cold-hardy cactus is perfect for high-elevation landscapes in the Southwest. Usually less than a foot tall, it tends to have few branches rather than the extensive mounds made by other species in the genus. Its diminutive size is compensated for by its spectacular blooms: Bright pink, saucer-shaped flowers appear in spring and summer, giving a pop of color against the dry grasses that Fendler's hedgehog often hides among. There is typically one stout central spine, with the others flattened against the stems in a star shape, making this a relatively easy hedgehog to handle when planting. At lower elevations, it may require some shade, but in upland gardens, this cactus will tolerate full sun and extreme cold, with the primary consideration being a well-draining soil that mimics the rocky sites that host this species in habitat. Ideal for crevice gardens or raised rock beds.

Echinocereus rigidissimus – Cactaceae

rainbow hedgehog cactus (cabeza de viejo)

5″–1′ × 3–5″

full sun, partial shade

low water needs

flowers May–July

Field guide: This resident of the sky islands along the US–Mexico border grows on steep rocky hillsides or sheer cliffs out of range of the hooves of grazing cattle. Look for it between 3500 and 6000 feet on granite and limestone outcrops and cliffs in Semidesert Grassland and Madrean Evergreen Woodland.

Description: Despite being a small plant that grows in obscure and hard-to-reach places, this is arguably one of the most distinctive cacti in the Southwest. It's an easy species to identify, with spines in bands of pink and white going up the stem, which is topped in late spring and early summer by pink flowers with cheery yellow centers. Most popular as a potted specimen, this species can also be mixed into a landscape if planted in well-draining soil. The spines are flattened against the plant, which makes for easy handling. Light shade from an overhanging mesquite (*Neltuma velutina*) will be beneficial at lower elevations, but this is an otherwise hardy plant that will add year-round visual interest to your yard and is sure to draw "oohs" and "ahhs" from the cactophiles in your life.

Escobaria vivipara (syn. *Pelecyphora vivipara*, *Coryphantha vivipara*) – Cactaceae

spinystar cactus (estrella de la tarde)

6″ × 6″–1′

full sun, partial shade

low water needs

flowers April–August

Field guide: Widespread from Mexico to Canada, bridging various habitats and growing conditions in its extensive range. In the Southwest, this plant is found between 2500 and 9000 feet in well-draining soil in almost every bioregion, excluding Sonoran Desertscrub communities and Subalpine Conifer Forest.

Description: One of the most cold-hardy cacti available to Southwestern gardeners, spinystar is a natural choice for high-elevation landscapes. Due to its wide distribution and numerous varieties, much taxonomic debate has surrounded this plant. In general, individuals of this species will form stout clumps of round heads the size and shape of golf or tennis balls, covered in a dense mat of spines splayed out into stars at the tips of evenly spaced, nipple-like protrusions. The bee-pollinated flowers of spinystar are light pink to rich fuchsia. Because of its small size, spinystar cactus benefits from being planted in groups, either scattered among low-growing wildflowers or set into a raised rock or crevice garden where the dainty but showy blooms will be easily visible.

Ferocactus acanthodes (syn. *Ferocactus cylindraceus*) – Cactaceae

California barrel cactus (biznaga)

4–5′ × 1–2′

full sun

low water needs

flowers February–June

Field guide: Grows on sandy wash edges, cliff faces, and gravelly slopes in Arizona Upland Sonoran Desertscrub, Lower Colorado River Sonoran Desertscrub, and Mojave Desertscrub between 500 and 4000 feet. You can find it on dry hillsides with creosote (*Larrea tridentata*) and chollas (*Cylindropuntia* spp.).

Description: This charismatic cactus species thrives in the hottest portions of the Sonoran and Mojave deserts, making it ideal for gardens exposed to full sun and reflected heat. This species is our tallest and most upright barrel cactus, covered in a thick filigree of scarlet-red spines that nearly obscure the plant's stem and protect it from the sun. The bright yellow flowers form a crown around the top of this cactus, blooming reliably even when moisture fails to materialize. For year-round visual interest and the ability to withstand the harshest conditions, California barrel is nearly unmatched among our native cacti.

Ferocactus wislizeni – Cactaceae

fishhook barrel cactus (biznaga de agua)

4–6′ × 2–3′

full sun

low water needs

flowers July–September

Field guide: Crosses the boundary between desert and upland habitats on rocky desert slopes in Arizona Upland Sonoran Desertscrub and Chihuahuan Desertscrub into Semidesert Grassland between 1000 and 4500 feet.

Description: This pill-shaped cactus has fiery red spines that curve into vicious hooks surrounded by whiskery radial spines. The flowers are typically glossy orange-red with numerous stamens emerging from the throat of each blossom and are primarily pollinated by chimney bees in the genus *Diadasia*; these bees will gorge themselves into a stupor in the pollen-rich blooms. Like many other plants, this cactus can produce nectar from structures other than its flowers, and this attracts ants that presumably protect the cactus's blooms from flower eaters that might damage the blooms. Plant this cactus in full sun or at the edge of a paloverde (*Parkinsonia* spp.) canopy and mix it in with other succulents such as agaves (*Agave* spp.) and saguaros (*Carnegiea gigantea*). They're also visually striking in mass plantings and perform well in relatively narrow areas like medians or sidewalk planters. Just remember to water sparingly and ensure good drainage.

Fouquieria splendens – Fouquieriaceae

ocotillo (ocotillo)

6–15′ × 6–12′

full sun

low water needs

flowers February–April, September–October

Field guide: This widespread succulent is found on sandy plains and gravelly slopes in Arizona Upland Sonoran Desertscrub, Lower Colorado River Sonoran Desertscrub, Mojave Desertscrub, and Chihuahuan Desertscrub into Semidesert Grassland at elevations ranging from 100 to 5000 feet. Not a hard plant to find in the desert southwest, it becomes even more common moving east from California toward Texas.

Description: Depending on the viewer's perspective, this plant's otherworldly appearance can be either alluring or repelling. Numerous sinewy, spine-covered branches with alternating rivulets of green and brown tissue reach up from a short trunk. Much of the year, these stems are leafless, but even slight rains will cause them to rapidly leaf out in spring and summer. They put on as much growth as possible before the soil dries up, at which point the leaves will turn shades of yellow or Vermont-in-fall red before dropping from the plant. The tubular red flowers appear in spring, regardless of rainfall, and bloom in seemingly perfect coordination with the migrations of several hummingbird species; ocotillo provides one of the most important and reliable sources of nectar for these winged jewels. Drought-tolerant and thriving on neglect, ocotillos are a must-have landscaping plant for gardeners in low-desert areas. Their ease of cultivation and essential role in attracting hummingbirds makes them ideal for beginners and experienced gardeners alike.

Grusonia emoryi – Cactaceae

devil's club cholla (casa de rata)

1′ × 3′+

full sun

low water needs

flowers May–June

Field guide: Thrives in sandy creosote flats, gravelly wash edges, and rocky slopes in Arizona Upland Sonoran Desertscrub and Chihuahuan Desertscrub from 2000 to 4500 feet. The population center of this species is found in Arizona between Safford and Globe, in the foothills of the mountains edging the Gila River.

Description: These much-neglected cousins of the better-known chollas and pricklypears deserve more recognition in Southwestern horticulture for their grotesque but intriguing forms and groundcover growth habit. They can be found across the deserts of the region, mostly in eastern Arizona and western New Mexico where rings form as the branches root into the ground and slowly migrate away from the center. The yellow flowers appear on the hottest days of early summer, followed by elongated yellow fruits that persist on the branch tips. Add this to your cactus garden, where its distinctive spreading form will differentiate it and contribute another texture and layer to your succulent flora.

Jatropha cardiophylla – Euphorbiaceae

limberbush (sangre de Cristo)

3′ × 2–3′

full sun

low water needs

flowers July–August

Field guide: Grows in a narrow north-south band running from near Tucson, Arizona, to the southern limits of the Sonoran Desert in Sonora, Mexico. Can be found between 1500 and 3500 feet on rocky, exposed slopes in Arizona Upland Sonoran Desertscrub, but is easiest to find during the monsoon season when its leaves are present.

Description: Limberbush bucks basically every standard traditionally used to define a good landscaping plant. Evergreen? Nope. Large, showy flowers? Certainly not. Poisonous? You bet! And yet this unusual plant really ought to be included in every Sonoran Desert garden. It bides its time through most of the year, waiting out cold and drought looking like nothing but a patch of dead sticks in the ground. But all at once, as rising humidity presages monsoon rains, the stems break out in an abundance of leathery, heart-shaped, lime-green leaves. As the rains begin to fall, white, urn-shaped flowers appear on thin stems that stick out just beyond the branch tips and are pollinated by a rare and enigmatic fly. This unusual plant's splash of bright green, a highlight of the monsoon season, is justification enough to include it in your landscape.

Lophocereus schottii – Cactaceae

senita (sinita)

6–15′ × 6–12′

full sun

low water needs

flowers March–September

Field guide: An important part of Sonoran Desert flora just south of the border, senitas reach the northern limit of their range in the Lower Colorado River and Arizona Upland subdivisions of the Sonoran Desert between 750 and 2000 feet.

Description: This cactus's multiple ribbed arms rise from the base to form an upright candelabra of spiny branches with stem tips that are often plum-purple. The lower spines are short and spreading, but as plants become reproductive, they develop long, whiskery spines at the tops of the branches. The small pink flowers are pollinated by a single moth species, the senita moth (*Upiga virescens*), and the females oviposit their eggs into the ovaries of senita flowers, where they develop as the flower turns into a plump red fruit. The resulting larvae eat their way out of the fruit and, next season, will mate on senita spines to start the process all over again. These extremely impressive specimens are particularly effective as vertical accents against west- or south-facing walls, where the reflected heat will protect them from frost.

Mammillaria grahamii (syn. *Cochemiea grahamii*) – Cactaceae

fishhook pincushion (cabeza de viejo)

6″ × 6″

full sun, partial shade

low water needs

flowers April–September

Field guide: This small cactus is found on rocky slopes and sandy wash edges in the Sonoran and Chihuahuan deserts in to Semidesert Grassland between 2000 and 5000 feet. You may pass by many individuals before noticing this plant, but once you know what to look for, you will see just how common it is.

Description: It's easy to miss this cactus at first glance, but when temperatures warm in spring, it goes into periodic fits of exuberant blooming with pink candy-striped flowers in a crown around the growth tip. This is followed by fruits that resemble tiny serrano peppers and taste a bit like kiwi. The diminutive stature of this cactus means that it is best suited to planting as a detail between larger, more charismatic species or wedged into rock gardens. Fishhook pincushion also lends itself well to small pots, where its clustering form and playful blooms will be more evident.

Nolina bigelovii – Asparagaceae

Bigelow's beargrass (yuca)

4–8′ × 4′

full sun

low water needs

flowers May–July

Field guide: This large succulent can be found on gravelly slopes and boulder fields between 1000 and 4000 feet. A denizen of the hot western edge of Arizona into southern Nevada, where it grows at the confluence of the Sonoran and Mojave deserts.

Description: One of the Southwest's most striking succulent plants, this species forms a shaggy ball of stiff, sharp-edged leaves topping a palm-like trunk thatched with dead foliage. The small but numerous flowers occur on a tall stalk protruding from the top of the plant, with pollen- and seed-bearing inflorescences arising from separate individuals. With its upright form and tolerance for reflected heat, it does well planted along a hot, exposed wall, where it will add a striking vertical element. Despite being razor-sharp and waxy, the foliage of this species is a larval food source for the gray hairstreak butterfly (*Strymon melinus*). Combine the blue-gray leaves of Bigelow's beargrass with the striking red spines of California barrel cactus (*Ferocactus acanthodes*) and darker green foliage of oneseed juniper (*Juniperus monosperma*) for a particularly effective planting.

Nolina microcarpa – Asparagaceae

beargrass (sacahuista)

4′ × 4′

full sun, partial shade

low water needs

flowers May–July

Field guide: Occurs on dry slopes across a wide range that spans Semidesert Grassland, Great Basin Grassland, Interior Chaparral, Madrean Evergreen Woodland, and Great Basin Conifer Woodland. Found between 3000 and 6500 feet in Arizona and New Mexico.

Description: This species forms clumps of thin, succulent leaves that resemble a large bunchgrass, though it is unrelated to true grasses. The leaves have sharp margins with minute serrations, and small, cream-colored flowers line thin, tall stalks arising from the rosette of leaves; they attract bees and pollinating beetles. Beargrass performs best in well-draining sandy or gravelly soils and does well on steep slopes, rock gardens, and even in large pots where the leaves can drape over the edges. In place of large tussock grasses in desert gardens, its evergreen nature ensures year-round visual interest, and its heat tolerance allows this succulent to thrive where true grasses would wither. Use as a foundation planting near buildings and walls, plant near the canopy edge of a tree surrounded with mounding perennials, or set it at the border of a raised rock garden with agaves and cacti.

Opuntia basilaris – Cactaceae

beavertail pricklypear (nopal)

1.5′ × 2–3′

full sun

low water needs

flowers February–June

Field guide: Found in southern Utah and Nevada and down the west side of Arizona, this plant grows on sandy wash edges and rocky slopes in Arizona Upland Sonoran Desertscrub, Lower Colorado River Sonoran Desertscrub, Great Basin Desertscrub, and Mojave Desertscrub into Interior Chaparral, Great Basin Grassland, and Great Basin Conifer Woodland. Occupies elevations as low as 1000 feet and as high as 5000 feet in warm microclimates.

Description: A distinctive sprawling pricklypear with blue-green to purple pads, this species is understandably popular in desert landscaping, where it stands out from the myriad cacti available to native plant gardeners. Its thick and succulent pads almost always lack spines but possess hair-like glochids that make handling the plant tricky; it's best done with tongs. The flowers are a stunning fuchsia, and the plants' early bloom season makes them one of the first to flower in native gardens. Often planted en masse along a walkway or in a highly visible front yard, this species does well alongside other desert plants like Bigelow's beargrass (*Nolina bigelovii*), oneseed juniper (*Juniperus monosperma*), and flattop buckwheat (*Eriogonum fasciculatum*).

Opuntia engelmannii – Cactaceae

Engelmann pricklypear (joconostle)

3–6′ × 5–7′

full sun, partial shade

low water needs

flowers April–July

Field guide: Grows between 1500 and 7500 feet on sandy flats and rocky slopes across almost all the bioregions of the Southwest, excluding the highest reaches of our mountain ranges.

Description: This widespread and variable species comprises an assortment of forms and hybrids that have spawned a litany of varietal names. The typical Engelmann pricklypear, if such exists, is a large, shrubby succulent with thick pads stacked three or more high, sporting white spines that are often in trios. The flowers are generally yellow all the way through, with no red striping near the bases. The variation in size and habitat preference of this plant means it is best to source material from relatively close to your garden to ensure that your pricklypear will be tolerant of local conditions. Its large size makes a strong statement in a garden, particularly among other cacti and agaves in a mixed-succulent bed. The flowers are an excellent nectar source for chimney bees, and the plump maroon fruits are favored by birds, desert tortoises, and humans for their juicy, semi-sweet pulp.

Opuntia polyacantha – Cactaceae

plains pricklypear

6″–1′ × 1–3′

full sun, partial shade

low water needs

flowers May–July

Field guide: Grows in well-draining soils in many Southwestern bioregions but is most prevalent in the Mojave Desertscrub and Great Basin bioregions (Conifer Woodland, Desertscrub, Grassland, and Montane Scrub) between 3000 and 8000 feet. One of only four cacti that occur naturally in Canada.

Description: This extremely cold-tolerant, mid- to high-elevation pricklypear can spend much of winter buried in snow with little negative effect. Its wide distribution equates to significant variation between populations and a spate of varieties. Most notable is the grizzlybear pricklypear (*O. polyacantha* var. *erinacea*), which is covered in a fur-like jacket of long, hairy spines and is found in the Mojave Desert. The more typical plains pricklypear is low and spreading with small pads rooting in where they touch the soil, eventually forming large rings along rocky slopes or half-buried in sand. It can be grown among other cold-tolerant shrubs and trees such as sagebrushes (*Artemisia* spp.) and junipers (*Juniperus* spp.) or planted on a succulent mound with banana yucca (*Yucca baccata*) and Fendler's hedgehog (*Echinocereus fendleri*).

Opuntia santa-rita – Cactaceae

Santa Rita pricklypear (nopal morado)

4′ × 4′

full sun, partial shade

low water needs

flowers April–June

Field guide: Found between 2500 and 4500 feet on rocky slopes and outcroppings in Semidesert Grassland and Madrean Evergreen Woodland in the sky islands spanning the border between Arizona and Sonora, Mexico.

Description: This plant's versatility and striking colors have brought it horticultural popularity that has helped spread it to gardens far away from its natural home in southeastern Arizona. Beloved for its paucity of spines and its rich purple color, this upright pricklypear grows well in low- to mid-elevation gardens in the desert Southwest. The flowers are bright yellow and pollinated by chimney bees. This species is sometimes visited by cochineal insects (*Dactylopius coccus*), which are sucking insects that leave behind a white film on plants (the film can easily be sprayed off using a hose). These insects' bodies are dried and ground to produce a bright red dye known as *carmine*, which has been used for millennia to dye fabrics and is still utilized today to make food coloring and lipstick.

Stenocereus thurberi – Cactaceae

organ pipe cactus (pitahaya dulce)

9–15′ × 9–12′

full sun, partial shade

low water needs

flowers April–July

Field guide: Limited in its US range, this plant is found between 1000 and 3000 feet on rocky desert slopes. The best place to see it is at Organ Pipe Cactus National Monument, a must-visit location in years when winter and spring rains encourage prolific wildflower displays.

Description: The relatively fast (compared to other columnar cacti) growth rate and beautiful form of this cactus make it common in the nursery trade. The main limiting factor for this desert species is cold—drought and heat are no problem—so plant it in warm locations with reflected heat, such as west-facing walls. The blooms form in early summer and open at night. The waxy-white tubular blossoms attract nectar-feeding bats and close as the heat of the day sets in. After a few years, this cactus will form a large candelabra of spiny branches. Once they've been cleaned of their spines, the fruits of organ pipe cacti are a particularly choice edible. This species can cover a large area when mature, so give it room to spread.

Yucca angustissima – Asparagaceae

narrowleaf yucca (palmilla)

2–3′ × 2–3′+ (can form colonies)

full sun

low water needs

flowers May–June

Field guide: Found mostly north of the Mogollon Rim on sandy plains or dunes, and on sandstone outcrops in Great Basin Desertscrub, Great Basin Montane Scrub, and Great Basin Conifer Woodland between 3000 and 7500 feet.

Description: This low-growing and cold-tolerant succulent thrives in Great Basin habitats, where it may form sizable colonies. The smaller stature of this species relative to other yuccas makes it suitable for more confined planting areas, such as courtyards, patios, or even pots. This species will not balk at full-sun exposure and can take reflected heat in mid- to high-elevation gardens, so there is no concern about placing it near a wall or other heat sink. The thin blue-green leaves are lined with fine white hairs that make decorative curlicue shapes, adding to the charm of this succulent. The flower stalk rises from the center of the plant, sporting showy purple buds that open into creamy white bell-shaped blossoms. Like all yuccas, this species is pollinated by female yucca moths (*Tegeticula* spp.), who forcibly insert their eggs into the ovary of the flower so that their larvae may hatch and grow inside the developing seed capsule. This relationship is one of the few known instances of intentional pollination of a flower by an insect.

Yucca baccata – Asparagaceae

banana yucca (dátil)

3–7′ × 3–9′+ (can form colonies)

full sun, partial shade

low water needs

flowers April–July

Field guide: Can be found in a wide swath of the Southwest from 3000 to 8000 feet, on rocky slopes and sandy flats in all bioregions except Lower Colorado River Sonoran Desertscrub and Subalpine Conifer Forest.

Description: Perhaps the most widespread Southwestern yucca, this species spans a range of habitats and can be quite variable from one population to the next. Generally, plants near the Mexican border (var. *brevifolia*) have upright trunks that can be several feet tall, while those farther north (var. *baccata*) are compact and low to the ground. Plants may form single rosettes or multibranched patches that can cover a hillside. Whatever the growth habit, banana yuccas are always armed with stiff, bayonet-like leaves; wear protective clothing and eyewear when working around this plant! Banana yuccas can handle a blanket of snow or baking in a strip-mall parking lot with no apparent ill effects. The flower stalks are short compared to those of other yuccas. The flowers are white, often streaked with pink on the outer petals, and they bear fruits that vaguely resemble bananas (if you squint). The fruits can be roasted, making a decent if not spectacular wild edible.

Yucca brevifolia – Asparagaceae

Joshua tree

15–30′ × 15–30′

full sun

low water needs

flowers March–May

Field guide: Joshua tree is the iconic species of the Mojave Desert, much as the saguaro is to the Sonoran. In western Arizona, along the Joshua Tree Parkway that connects Phoenix to Las Vegas, these two deserts meet, and their two keystone species mingle. Look for Joshua trees on sandy flats and gravelly slopes between 2500 and 3500 feet in western Arizona, southern Nevada, Utah, and eastern California.

Description: Joshua trees are ungainly but beautiful, with thin trunks and mops of blue-green succulent leaves that seem lifted right out of a Dr. Seuss book. Often growing to impressive size, with dozens of twisting arms weaving into one another or snaking through the air, these are statement plants with the potential to elevate an ordinary garden to a whimsical one. Plant in well-draining soil and give your Joshua tree plenty of room to reach its full size without danger to nearby structures or parked cars.

Yucca elata – Asparagaceae

soaptree yucca (cortadillo)

6–20′ × 3–15′

full sun

low water needs

flowers May–July

Field guide: Found on sandy plains and rocky slopes in Arizona Upland Sonoran Desertscrub, Chihuahuan Desertscrub, and Great Basin Desertscrub into Semidesert Grassland and Great Basin Grassland, where it is found between 2500 and 6500 feet.

Description: This tree-like succulent towers over the grasslands of the Southwest. It is ideal for mid-elevation gardens and can be used to anchor a garden bed, surrounded by grasses, shrubs, and wildflowers. The pliable leaves are blue with white margins and fine curly hairs. Over time, the lower leaves will die but stay attached to the plant. It is important to leave this thatch of foliage, both to protect the plant from sun and frost and to maintain the distinctive aesthetic that makes this species so desirable as a landscape element. Soaptree yuccas can be very long-lived, with some individuals reaching well over 200 years, so this plant may enhance the aesthetic of your garden for decades or even centuries to come!

Yucca schottii (syn. *Yucca madrensis*) – Asparagaceae

mountain yucca (soco)

6–9′ × 3–6′

full sun, partial shade

low water needs

flowers July–August

Field guide: This species is only found in the sky islands along the international boundary, where it grows in Semi-desert Grassland and Madrean Evergreen Woodland at elevations of 4000 to 6500 feet.

Description: From a landscaping perspective, one of the interesting aspects of this species is its ability to handle shade, something that most yuccas won't tolerate. Mountain yuccas have long, sharp-tipped leaves that may be stiff or somewhat pliable but are always a pain to run into. The flowers are dense on the stalks, forming an uninterrupted mass of cream-colored blossoms that turn into multi-chambered capsules that split and drop flat, black seeds. Use mountain yucca in a garden that already has mature shade trees or where a low-water plant is desired in an area that might only get sun through part of the day.

Grasses

Grasses are often underappreciated and underutilized in landscapes, an unfortunate fact that extends from the beginner gardener to the experienced landscaper. Less charismatic than showy wildflowers or distinctive succulents, grasses are excluded from garden designs entirely or, worse, shunned in favor of non-native species. Grasses are integral to the ecosystems of most of the Southwest, occurring at all elevations and across the whole geography of the region, with the grass family (*Poaceae*) constituting one of the three largest plant families in our region. The judicious use of native grasses in a garden elevates it visually and ecologically by integrating the wiry shape of these plants and the way they sway in a breeze to add a sense of motion to the landscape. Grasses may not seem like important pollinator plants since they lack nectar-rich blossoms, but their leaves are a larval food source for a whole host of skipper butterflies (*Hesperiidae* spp.), and the foliage provides cover for wildlife ranging from insects to birds and mammals. Grasses are also an important food source for desert tortoises (*Gopherus* spp.), making them an essential component of any garden with a resident tortoise. The fibrous root system of grasses makes them ideal for erosion control, well suited for planting in basins or swales and breaking up heavy clays to pave the way for shrubs and trees.

Achnatherum hymenoides (syn. *Eriocoma hymenoides*) – Poaceae

ricegrass (arroz Indio)

3′ × 2′

full sun, partial shade

low water needs

flowers May–August

Field guide: Found between 3500 and 6500 feet, north of the Mogollon Rim that cuts across central Arizona and southern New Mexico. One of our most widespread upland grass species, especially common in Great Basin Grassland and Great Basin Conifer Woodland into Montane Conifer Forest. This is the state grass of Nevada.

Description: Most of our native grasses are warm-season species that green up only in the hot, humid months of midsummer through early fall. This delightful exception lends its lacy texture and ornate inflorescence to a garden early in the year, as trees begin to leaf out and spring wildflowers come into bloom. The seeds of this plant have been a food source for people in the Southwest for millennia and are relished by wildlife, including birds and small mammals. The roots of this grass seem to be associated with increased microbial activity and form a structure known as a *rhizosheath* that helps them cope with aridity and heat in the nutrient-poor sandy soils they naturally occupy. This suggests that ricegrass may help improve soil quality by partnering with microorganisms in a manner similar to legumes like velvet mesquite (*Neltuma velutina*) and New Mexico locust (*Robinia neomexicana*), a trait that should earn ricegrass a spot in any water-wise high-elevation Southwestern garden.

Aristida purpurea – Poaceae

purple threeawn

1.5′ × 1′

full sun

low water needs

flowers April–October

Field guide: Incredibly widespread in the Southwest, it is probably safe to say this plant is found in every bioregion with the exclusion of Subalpine Conifer Forest. Grows in gravelly and sandy soils on slopes or flats between 1000 and 7000 feet, especially around disturbed sites.

Description: This widespread shortgrass is distinctive for its nodding form and the bright pink-purple hue of its inflorescence. A versatile plant, it will be equally at home in a cactus garden with saguaro (*Carnegiea gigantea*) and fishhook barrel (*Ferocactus wislizeni*) or a meadow planting with tufted evening primrose (*Oenothera cespitosa*), blue flax (*Linum lewisii*), and Mexican hat (*Ratibida columnifera*). One stand-out trait is that it tends to green up with both winter and summer rains, which means that it sports its eye-catching purple inflorescence for much of the year. The seeds readily disperse and germinate in open spaces, so this plant will volunteer along paths, in basins, and tucked into rockwork—a trait to be encouraged, as the fibrous roots help bind and enrich soil.

Aristida ternipes – Poaceae

spidergrass (zacate araña)

3′ × 2′

full sun

low water needs

flowers April–October

Field guide: Grows between 2500 to 5500 feet from the Sonoran and Mojave deserts into Semidesert Grassland. This species is often one of the first to appear in recently disturbed areas.

Description: The attractive wiry shape of the inflorescence and the extreme hardiness of this species make it a good choice for any low- to mid-elevation landscape. Spidergrass is a reliable reseeder due to the inflorescence's ability to detach from the plant and blow around like a tiny tumbleweed, a trait which makes it ideal for disturbed sites or restoration areas where plant biomass is needed quickly to restore soil and add organic matter. This plant is best set away from path edges where its pointy awns (needle-like appendages attached to the spikelets) might catch on the clothing, fur, or skin of passersby.

Blepharoneuron tricholepis
(syn. *Muhlenbergia tricholepis*) – Poaceae

hairy dropseed (popotillo del pinar)

2.5′ × 2′

full sun, partial shade

low water needs

flowers June–November

Field guide: Grows in mountain meadows, clearings, and roadsides in Montane Conifer Forest and Subalpine Conifer Forest from 6000 to 12,000 feet.

Description: This species is one of our finest grasses for high-elevation gardens, with hairy, silver-pink spikelets that remain on the plant as it dries to a straw color in fall. The dense rosette of leaves and attractive inflorescence look particularly good in a mass planting under ponderosa pine (*Pinus ponderosa*) or white fir (*Abies concolor*), interspersed with cold-tolerant wildflowers like pineywoods geranium (*Geranium caespitosum*) and western wallflower (*Erysimum capitatum*). Birds and small mammals will eat the seeds of this species, and though it does best in open spaces, it will tolerate some shade and can grow in north-facing exposures.

Bothriochloa barbinodis – Poaceae

cane beardgrass (zacate popotillo)

3–4′ × 2–3′

full sun

low water needs

flowers April–October

Field guide: Can be found in all the Southwest's deserts and into grassland and woodland habitats from 1500 to 6000 feet. Throughout its range, it tends to appear following disturbance events and fades out as larger shrubs and trees create a shady overstory.

Description: This tough, versatile, and aggressive grass quickly spreads into open spaces, from rolling grasslands to sidewalk cracks. The dense, silky hairs and pointy awns act as good identifiers for this common species and make it appealing in a landscape. In addition, the lime-green leaves turn a rusty red in fall and add color and texture to landscapes beginning to go dormant for the cool season. Because it can readily reseed, this species is liable to pop up around the yard and can become a nuisance if planted into beds already densely populated with other plants; few things are less fun than trying to weed cane beardgrass out from under cholla! However, this plant's aggressive habit can be a benefit in vacant yards or restoration sites where fast growth and a robust root system are called for. It is particularly effective at preventing erosion in basins and on hillsides.

Bouteloua chondrosioides – Poaceae

sprucetop grama (navajita morada)

2′ × 1′

full sun

low water needs

flowers August–October

Field guide: This grass can be found on open plains and rocky slopes from 2500 to 6000 feet in Semidesert Grassland and Madrean Evergreen Woodland habitats.

Description: This charming shortgrass has banner-like spikelets, which turn a dark cherry-red and sport dangling orange anthers that defy anyone who would question the beauty of grass flowers. The species forms small tight clumps and stands out in a landscape when planted with a mix of other grasses and wildflowers. Growing best on thin-soiled slopes in full sun, sprucetop grama makes a good terrace or basin-edge planting where its fibrous root system will help bind the soil.

Bouteloua curtipendula – Poaceae

sideoats grama (navajita banderilla)

2–3′ × 1–2′

full sun, partial shade

low water needs

flowers June–November

Field guide: This widespread species's range encompasses all four Southwestern deserts and extends into Madrean Evergreen Woodland, Great Basin Conifer Woodland, and Montane Conifer Forest habitats. Look for it from 2500 to 7000 feet on slopes, plains, rocky outcrops, and in woodland meadows.

Description: One of the most ubiquitous and recognizable grasses of the Southwest, this is a fantastic landscape grass for beginners and native gardening experts alike. Identifiable by its upright habit and neat rows of spikelet-bearing branches dangling from the inflorescence like flags from a pole, it tolerates well-watered basins and swales, minimally irrigated desert gardens, or dense prairie plantings with equal aplomb and reliability. A highly palatable forage species for tortoises, this plant's seeds are sought after by birds like lesser goldfinches (*Spinus psaltria*), verdins (*Auriparus flaviceps*), and Gambel's quail (*Callipepla gambelii*), and the foliage is a larval food source for checkerspot butterflies (*Euphydryas* spp.) as well as the veined ctenucha moth (*Ctenucha venosa*). Best of all, unlike many of our other native grasses, sideoats grama is commonly available in most local nurseries.

Bouteloua gracilis – Poaceae

blue grama (navajita común)

1.5–2′ × 1′

full sun

medium water needs

flowers July–October

Field guide: In grasslands, woodlands, and mountain meadows from Utah and Colorado to the Mexican border, this species intersperses itself among other grasses, perennials, and shrubs from 4000 to 8000 feet. It is common where fire or grazing has removed taller grass species that would crowd it out.

Description: Blue grama is relatively short in stature, but it punches above its weight class in terms of visual appeal. The inflorescences rise above the bluish-green foliage and hold spikelets that resemble eyelashes that curl into spirals as they age. Plant blue grama in a shallow basin or swale with scattered shrubs and flowers such as littleleaf sumac (*Rhus microphylla*) and blackfoot daisy (*Melampodium leucanthum*), or in a prairie planting with other grasses such as sideoats grama (*Bouteloua curtipendula*), plains lovegrass (*Eragrostis intermedia*), and little bluestem (*Schizachyrium scoparium*).

Bromus carinatus – Poaceae

Arizona brome (bromo)

2.5′ × 2′

full sun, partial shade, shade

medium to high water needs

flowers July–November

Field guide: This species is often found in the dappled light of Madrean Evergreen Woodland, Great Basin Conifer Woodland, or Montane Conifer Forest habitats. Look for it in moist roadside ditches or stream margins from 3500 to 9000 feet.

Description: Arizona brome is an elegant high-elevation grass fond of wet spots like basins and gray-water outlets. Give this species regular moisture during the hot months and plant it in some shade, either under a tree or on the north side of a structure. The elongated spikelets and nodding inflorescence of this species give it a graceful appearance that is highlighted when it's planted under the shade of Arizona white oak (*Quercus arizonica*) or netleaf hackberry (*Celtis reticulata*) and mixed in with showy fleabane (*Erigeron speciosus*), Hooker's evening primrose (*Oenothera elata*), and Hartweg's sundrops (*Calylophus hartwegii*).

Digitaria californica – Poaceae

Arizona cottontop (zacate punta blanca)

3′ × 2′

full sun, partial shade

medium water needs

flowers August–October

Field guide: Found from just over 1000 feet in Lower Colorado River Sonoran Desertscrub near Yuma, Arizona, to almost 6000 feet in the Great Basin Conifer Woodland north of Silver City, New Mexico, this species likes rocky slopes and outcrops, where it often appears following disturbances (such as fire).

Description: This plant's distinctive fuzzy inflorescence glows in the rays of the rising or setting sun, making it a particularly decorative grass for Southwestern landscapes. It establishes quickly and is perfect for new gardens where larger trees and shrubs have yet to establish, and it will tend to fade out as other plants crowd out the canopy and root space. Forming a tight clump of bright foliage, Arizona cottontop can be used as a mass planting to cover a bare slope, or it can be scattered in with wildflowers and grasses in a pollinator and wildlife habitat bed. Its fast growth rate makes it an ideal desert-tortoise food source.

Disakisperma dubium – Poaceae

green sprangletop (zacate gigante)

4′ × 3′

full sun, partial shade

medium water needs

flowers July–October

Field guide: Grows in washes, drainages, and meadows from 2500 to 6000 feet in Arizona Upland Sonoran Desertscrub, Chihuahuan Desertscrub, Semidesert Grassland, and Madrean Evergreen Woodland habitats. The range of this species is primarily south of I-40, in the southern portions of Arizona and New Mexico and into west Texas.

Description: This large, precocious grass can reach several feet in height, with an inflorescence that sits at the top of long stems and has an untidy, windblown-hair look to it. Birds eat the seeds, and the foliage attracts checkerspot butterfly (*Euphydryas* spp.) caterpillars and serves as an abundant food source for desert tortoises. Because this grass likes a little extra water, it is best planted in a basin or near a gray-water outlet where it can receive the moisture it needs to thrive. Try planting with other species that have higher water needs, like horsetail milkweed (*Asclepias subverticillata*) and seep willow (*Baccharis salicifolia*).

Elionurus barbiculmis – Poaceae

woolyspike balsamscale

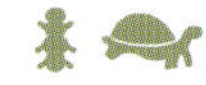

2.5′ × 2′

full sun, partial shade

low water needs

flowers July–October

Field guide: This grass occurs in the sky islands of southern Arizona and New Mexico, where it is the northernmost representative of a largely tropical grass genus. Look for it from 4000 to 6000 feet on open slopes and canyon edges in Semidesert Grassland and Madrean Evergreen Woodland.

Description: Though this species is not yet common in the nursery trade, its decorative scaled inflorescence and neat sprays of rolled leaves should warrant wider use in landscapes. Woolyspike balsamscale prefers well-draining rocky soils and is likely to languish in heavy clays. In low-elevation gardens, it may benefit from being planted where it will receive afternoon shade or filtered light from an overhanging velvet mesquite (*Neltuma velutina*) or netleaf hackberry (*Celtis reticulata*). In upland gardens, full sun is appropriate, and this grass makes a nice companion planting with desert spoon (*Dasylirion wheeleri*) and pointleaf manzanita (*Arctostaphylos pungens*).

Elymus trachycaulus – Poaceae

slender wheatgrass

3′ × 2′

full sun, partial shade

medium water needs

flowers July–October

Field guide: Found in sandy bottomlands and woodland clearings from 3500 to 9000 feet across most of the American West, where it is particularly common in Great Basin Grassland and Great Basin Conifer Woodland settings.

Description: Slender wheatgrass is a distinctive species with ashy blue foliage and an inflorescence that consists of an attractive spike that dries from blue to tan in fall. This species forms clumps, sometimes connected by stout underground runners that make it useful for holding together soil in drainages, basins, and swales. The closely related western wheatgrass (*Pascopyrum smithii*) has a similar aesthetic and spreads by wide-ranging runners that also make it ideal for erosion control. Either of these species will work in mid- to high-elevation gardens where the foliage color will contrast nicely against darker green shrubs and grasses.

Eragrostis intermedia – Poaceae

plains lovegrass (zacate llanero)

3′ × 2′

full sun, partial shade

low water needs

flowers June–October

Field guide: Find this grass from 3000 to 6000 feet in open fields and woodland edges in Semidesert Grassland, Great Basin Grassland, Madrean Evergreen Woodland, and Great Basin Conifer Woodland. Particularly abundant stands of this species can be found in the sky islands of southeastern Arizona.

Description: The genus *Eragrostis* has a nasty reputation in the Southwest for having supplied several of our most pernicious invasive grass species: Lehmann lovegrass (*Eragrostis lehmanniana*), weeping lovegrass (*Eragrostis curvula*), and Wilman lovegrass (*Eragrostis superba*). Yet this same genus has also given us this native species that is ideal for restoration and landscaping. The open, diffuse panicles of this plant create a pink mist over the ground in summer, and the teardrop-shaped spikelets provide food for native birds like Gambel's and scaled quail (*Callipepla gambelii* and *C. squamata*). It thrives in well-draining sandy or loamy soil, so in clay areas it should be planted in soil amended to improve drainage. Combine with bullgrass (*Muhlenbergia emersleyi*), green sprangletop (*Disakisperma dubium*), and blue grama (*Bouteloua gracilis*) for a texturally diverse grassland garden, then drop in velvetpod mimosa (*Mimosa dysocarpa*) and Goodding's mock vervain (*Glandularia gooddingii*) for pops of color.

Festuca arizonica – Poaceae

Arizona fescue

2–3′ × 2–3′

full sun, partial shade

medium to high water needs

flowers June–August

Field guide: Thrives in understories and open meadows in Montane Conifer Forest and Subalpine Conifer Forest from 6000 to 10,000 feet.

Description: A gorgeous fountain of green-blue foliage and nodding inflorescences, this is one of the best landscaping grasses for high-elevation gardens. Its fibrous root system helps bind soil, and studies have shown that it hosts a diverse assemblage of mycorrhizae that help enrich garden soils over time while improving the drought tolerance of the plant. Individuals are known to live up to twenty years in habitat, making this species a stable and reliable part of your landscape. Arizona fescue is often grown in mass plantings along walls or pathways or mixed into a garden with wildflowers such as Hooker's evening primrose (*Oenothera elata*) and pineywoods geranium (*Geranium caespitosum*). Occasionally combing out dead foliage or cutting the plant back every one to two years will ensure abundant fresh growth.

Hesperostipa neomexicana – Poaceae

New Mexico feathergrass

3′ × 2′

full sun

low to medium water needs

flowers April–June

Field guide: Found in Great Basin Grassland, Interior Chaparral, and Great Basin Conifer Woodland habitats from 3500 to 6500 feet, where it is often seen on sandy plains, gravelly slopes, or rock outcrops.

Description: This beautiful and distinctive native species is woefully underutilized in landscaping. It features whip-like plumed awns that twist and curl like calligraphy letters and catch sunlight as they bob above the wispy sprays of foliage. The seeds will readily germinate where they fall when rain or irrigation are abundant enough. A drought-tolerant grass, this species can be mixed in with cacti and succulents as well as shrubs and wildflowers. New Mexico feathergrass is not to be confused with Mexican feathergrass (*Nassella tenuissima*), a lovely species that is also native to the Southwest but has become invasive in California, which shows how even plants that evolved just a few hundred miles away can become problematic when moved into new areas.

Heteropogon contortus – Poaceae

tanglehead
(zacate colorado)

3′ × 2′

full sun

low water needs

flowers August–October

Field guide: This common grass species thrives in disturbed areas from 1000 feet in Sonoran Desert washes to 5500 feet in grazed areas and trail margins in Madrean Evergreen Woodland. Primarily found in southern Arizona and New Mexico, it also occurs around the Grand Canyon.

Description: Easily distinguished from other grasses by its dark inflorescence with long awns that twist together and detach in a cluster, often sticking to pant legs or forming tangled masses on the ground, this grass is quite drought-tolerant. Its ability to spread by seed can make it challenging to control in a mixed-grass planting, especially in smaller gardens; this trait and the plant's tolerance for disturbed sites make it an ideal species for restoration projects or filling large areas. When in flower, it's often swarming with butterflies, bees, beetles, wasps, and ants. This is unusual for a grass and occurs when the inflorescences become infected by a smut fungus (*Sporisorium* spp.), which releases a sticky liquid irresistible to insects. Additionally, some skipper butterflies will utilize the foliage as a caterpillar food source. Tanglehead is one of our hardiest and most decorative grasses, with potential to be as good a pollinator attractor as any wildflower in your garden (if you have space for it).

Hilaria belangeri – Poaceae

curly mesquite grass (zacate chino)

9″ × 3–6′

full sun

low water needs

flowers July–November

Field guide: Found in Semidesert Grassland on thin-soiled slopes and rock outcrops, or recently disturbed hillsides and washes, from 3000 to 6000 feet.

Description: The small stature of this species is no reason to ignore it as a landscaping selection. This mat-forming plant spreads by aboveground runners called *stolons* (this is also how strawberries spread) that knit together disturbed soil, making it a good choice for restoration and erosion control. At dawn, especially during the monsoon season, it is not unheard of to see butterflies hanging around the leaves of this and other grasses, lapping up the nutrient-infused water that seeps from the foliage in a process called *guttation*. Plant curly mesquite grass along eroding slopes, as a drought-tolerant turf lining pathways, or as a groundcover and green mulch in a bed of wildflowers.

Hilaria rigida – Poaceae

big galleta (toboso)

3′ × 4′

full sun

low water needs

flowers February–November

Field guide: Look for this species on desiccated flats, rocky slopes, and barren dunes across the Sonoran and Mojave deserts, from nearly sea level to 4000 feet.

Description: This stout, shrubby grass successfully grows in some of the harshest and hottest portions of the Southwest. A bushy species, it spreads to form dense clumps of stiff leaves and narrow spikes of seeds, which reveal the zig-zag pattern of the inflorescence once they fall. Though significant dieback may occur during periods of extended drought, this plant readily regrows following rains, and occasional combing or trimming will help it maintain a fresh look in the garden. It will be right at home in a low-water garden with creosote bush (*Larrea tridentata*) and Joshua tree (*Yucca brevifolia*), and seeding around annuals like bajada lupine (*Lupinus concinnus*), Mexican gold poppy (*Eschscholzia californica* var. *mexicana*), and purple owl clover (*Castilleja exserta*) will make for a stunning spring show, offsetting the cool-season dormancy of this grass.

Hopia obtusa – Poaceae

vine mesquite grass (zacate guía)

2.5′ × 4′+

full sun, partial shade, shade

medium water needs

flowers May–October

Field guide: Often grows in swales, streambeds, and drainages, where seasonal flooding provides the extra moisture it needs to thrive. Found as low as 2000 feet in the Sonoran Desert and as high as 6000 feet in Madrean Evergreen Woodland and Great Basin Conifer Woodland.

Description: Despite its common name, this species is neither a vine nor a mesquite. However, it does spread by above- and belowground runners, forming a thick groundcover that is often seen in the shade of mesquites (*Neltuma* spp.), oaks (*Quercus* spp.), or velvet ash (*Fraxinus velutina*). The large seeds are an excellent food source for birds and small mammals, while the prolific foliage and roots adeptly prevent erosion. This grass is best planted in the bottom of a basin where its roots can knit together the soil. Plant a tree on the outer edge to provide shade, and direct rainwater or gray water into the basin to ensure optimal growth of this lush and wildlife-friendly species.

Muhlenbergia alopecuroides
(syn. *Lycurus setosus*) – Poaceae

bristly wolfstail (zacate lobero)

2′ × 1.5′

full sun, partial shade

low water needs

flowers July–October

Field guide: Found on thin-soiled rocky slopes and open mesas in both Semidesert and Great Basin Grassland up through Montane Conifer Forest from 3000 to 8000 feet.

Description: Bristly wolfstail is a small but hardy bunchgrass that is tolerant of a range of soils and elevations, making it appropriate for almost any Southwestern landscape. The seeds are relished by chipping sparrows. Plant as part of a mixed grassland assemblage with blue grama (*Bouteloua gracilis*), plains lovegrass (*Eragrostis intermedia*), and prairie acacia (*Acaciella angustissima*). The spike-like inflorescence and compact form give this plant a distinctive look that helps it stand out from other grass species in a garden.

Muhlenbergia dumosa – Poaceae

bamboo muhly (carricillo)

4–5′ × 5–6′

full sun, partial shade

low water needs

flowers January–May

Field guide: Can be found from 2500 feet at the upper limits of the Sonoran Desert through Semidesert Grassland and into Madrean Evergreen Woodland at 5500 feet. This sky-island species grows on canyon walls and rocky slopes along the Arizona–Sonora boundary.

Description: True to its name, this species has numerous stems arising from a spreading root system, giving it the look of a dense grove of miniature bamboo canes. This growth habit makes it useful as an informal hedge or as a foundation planting along walls, where its swaying stems will cast intricate shadow patterns. Birds such as lesser goldfinches (*Spinus psaltria*) and verdins (*Auriparus flaviceps*) relish the seeds, and bees, butterflies, and other insects use the thicket of stems as nesting or pupating sites. When well-watered, this grass will retain green growth year-round, contributing visual interest to winter landscapes. It prefers well-draining soils but is not fussy about soil types or exposures and is equally at home in a succulent garden with Parry's agave (*Agave parryi*) and ocotillo (*Fouquieria splendens*) as in a gray water–fed planting with netleaf hackberry (*Celtis reticulata*) and seep willow (*Baccharis salicifolia*).

Muhlenbergia emersleyi – Poaceae

bullgrass (zacate de toro)

4′ × 3′

full sun, partial shade

low water needs

flowers June–November

Field guide: Look for this species on open slopes, woodland edges, and canyon margins between 3500 and 6500 feet from Semidesert Grassland into Madrean Evergreen Woodland and Great Basin Conifer Woodland.

Description: One of the most widely planted native grasses and readily available in the nursery trade, this species is becoming increasingly common in public plantings and home gardens. Its popularity is easy to explain when summer rains bring forth tall, pink-hued panicles of tiny spikelets out of thick fountains of lush green foliage. That robust, dense foliage provides both forage and shelter for desert tortoises, nesting sites for native insects, and caterpillar food for the red-bordered satyr butterfly (*Gyrocheilus patrobas*), and the seeds are a food source for songbirds. The relatively large size of this grass is matched by its deep root system, which is highly effective at holding soil together and providing habitat for soil-dwelling organisms. It can be used as a foundation planting against a wall, to edge a pathway, or to fill a raised planter.

Muhlenbergia montana – Poaceae

mountain muhly

2.5′ × 2′

full sun, partial shade

low water needs

flowers July–November

Field guide: Where fires or rocky outcroppings cause openings in Great Basin Conifer Woodland, Montane Conifer Forest, or Subalpine Conifer Forest, this species tends to appear, taking advantage of a lack of competition from large trees for sunlight and moisture. Look for this species from 4500 to 10,000 feet.

Description: A wiry grass with purple-tinted spikelets tipped by hair-like awns, this species can be mixed in with high-elevation wildflowers or used to fill in spaces between trees. Mountain muhly's ability to grow on rocky outcrops also makes it a suitable choice for rock gardens and terraced beds where it can reseed and spread. This species is incredibly cold-tolerant and ideal for almost any high-elevation garden, but it will suffer in desert landscapes where bush muhly (*Muhlenbergia porteri*) or bullgrass (*Muhlenbergia emersleyi*) would be better choices.

Muhlenbergia porteri – Poaceae

bush muhly (zacate aparejo)

1.5′ × 3′+

full sun, partial shade

low water needs

flowers August–November

Field guide: Able to tolerate harsh desert flats and sun-baked rocky slopes, this species is often found tucked under shrubs and small trees, where it receives shade and protection from grazing animals. Grows from 2000 to 6000 feet across the Southwest.

Description: One of the hardiest perennial grasses in the Southwest, bush muhly is ideal for low-desert landscapes. This is one of our few native grasses that branches above the base, forming a dense tangle of stems that terminate in airy, open panicles partially enclosed by leaf sheaths. When in flower, this species resembles a pink mist over the ground and will readily reseed to fill in blank spaces in a garden. Mimic the natural tendency of this species to associate with overhanging shrubs or trees by setting it at the base of a whitethorn acacia (*Vachellia constricta*) or velvet mesquite (*Neltuma velutina*).

Muhlenbergia rigens – Poaceae

deergrass (zacate de venado)

3′ × 3′

full sun, partial shade

medium to high water needs

flowers July–October

Field guide: Occurs from 3000 to 8500 feet in Riparian Corridors, ranging from sandy desert washes to shaded canyons in Montane Conifer Forest.

Description: This is a strikingly lovely and graceful native grass that, in habitat, forms dense patches that slow and disperse the flow of water, helping prevent erosion while knitting together the soil. The affinity of deergrass for moist places makes it an excellent selection for basins, swales, and gray-water outlets where the extra moisture will encourage lush, robust sprays of foliage. The inflorescence resembles a blond spike jutting above the foliage, with spikelets massed close to the main stem. A well-balanced basin planting could consist of this species along with desert honeysuckle (*Anisacanthus thurberi*), Texas mulberry (*Morus microphylla*), and an overstory of netleaf hackberry (*Celtis reticulata*).

Pappostipa speciosa (syn. *Stipa speciosa*) – Poaceae

desert needlegrass (coirón)

2′ × 2′

full sun

low water needs

flowers March–June

Field guide: Gravitates to rocky desert hillsides and boulder-strewn Interior Chaparral in Arizona, southern Utah, and southwestern Colorado from 2000 to 6000 feet.

Description: Desert needlegrass has dense, silvery hairs on the spikelets that catch morning and evening light and impart a lovely glow to the tops of these plants. This species is both cold- and drought-tolerant and thrives in sandy, well-draining soils. Plant with oneseed juniper (*Juniperus monosperma*), sugar sumac (*Rhus ovata*), and Joshua tree (*Yucca brevifolia*) for a naturalistic plant palette (in other words, a grouping that closely resembles naturally occurring guilds).

Piptochaetium fimbriatum – Poaceae

pinyon ricegrass

2′ × 2′

partial shade, shade

medium water needs

flowers July–September

Field guide: Look for this in the woodlands and forests of southern New Mexico and Arizona, where it thrives in dappled light along canyon margins and slopes from 3500 to 7000 feet.

Description: This species has a languid, shaggy form with elegant, long-awned spikelets that make it one of our most decorative native grasses. A tolerance for shade means this plant will be appropriate for those hard-to-plant, low-light areas (such as on the north side of a building or in the deep shade of a mature tree where other species suffer). A gorgeous woodland guild can be made by planting a patch of pinyon ricegrass alongside Wright's silktassel (*Garrya wrightii*) and three-leaf sumac (*Rhus aromatica* var. *trilobata*) in the shade of an Emory oak (*Quercus emoryi*).

Schizachyrium scoparium – Poaceae

little bluestem

4′ × 2′

full sun, partial shade

low water needs

flowers July–October

Field guide: This species is a vital component of the Great Plains and Colorado Plateau grasslands but becomes sparser in southern Arizona and New Mexico, where it begins to be replaced by the similar crimson bluestem (*Schizachyrium sanguineum*). Look for this plant from 4000 to 8000 feet on open plains and in woodland clearings.

Description: A highly ornamental grass with blue-green stems that dry to a rusty red and are topped by spikelets covered in tufts of cottony hairs. The red fall color of this species contributes to a striking palette when combined with the gray foliage of winterfat (*Krascheninnikovia lanata*) and the green and red of three-leaf sumac (*Rhus aromatica* var. *trilobata*). The dried stems and seeds also make lovely additions to flower arrangements with flattop buckwheat (*Eriogonum fasciculatum*) and globemallow (*Sphaeralcea* spp.).

Setaria leucopila – Poaceae

plains bristlegrass (zacate tempranero)

3′ × 2′

full sun, partial shade

low water needs

flowers May–October

Field guide: This species grows in bottomlands and disturbed sites as low as 3000 feet in Arizona Upland Sonoran Desertscrub and Chihuahuan Desertscrub to over 5500 feet in Madrean Evergreen Woodland.

Description: This is a fantastic addition to a garden for anyone hoping to provide food for birds in their yard. The large, bristly seeds are a draw for a variety of avian species, who will eat them and kindly do the work of spreading them around your yard, with seedlings tending to come up under trees and along fence lines. This grass is sometimes mistaken for invasive species like buffelgrass (*Cenchrus ciliaris*) and fountaingrass (*Cenchrus setaceus*) by overenthusiastic but underinformed plant vigilantes. However, along with large-spike bristlegrass (*Setaria macrostachya*), this is one of our most valuable native grasses for wildlife.

Sporobolus wrightii – Poaceae

giant sacaton (zacatón)

5–6′ × 5′

full sun, partial shade

medium water needs

flowers March–November

Field guide: Historically, this species dominated waterways and moist lowlands around the Southwest but is now estimated to occur in only 5 percent of its former range. It can be seen from 2000 to 6500 feet in Riparian Corridors. In salty flats, this species is replaced by the similar but smaller alkali sacaton (*Sporobolus airoides*).

Description: The most stately of our native grasses, with foot-long pyramidal inflorescences rising out of luxurious thickets of foliage. Its impressive size means it functions well as a focal specimen or accent planting and can even be used as an informal hedge. Because of sacaton's affinity for wet spaces, consider planting it in basins or gray-water outlets where it will receive the moisture it needs to thrive. It is an excellent desert tortoise food and shelter plant and a nesting site for Botteri's sparrow (*Peucaea botterii*). Granivorous birds eat the seeds, which have also served as a protein-rich food source for human residents of the Southwest for millennia.

Zuloagaea bulbosa – Poaceae

bulb panicgrass (panizo de bulbo)

5–6′ × 3–4′

full sun, partial shade

medium water needs

flowers July–October

Field guide: This species thrives in wetlands, drainages, and roadside ditches from 4000 to 8000 feet, in habitats ranging from Semidesert Grassland to Montane Conifer Forest.

Description: This upright species has easily harvested seeds that have long been relished by the wildlife and people of the Southwest. It's an interesting grass: from the diffuse branches of the inflorescence, covered in pink spikelets that nod and bow in the breeze, to the shallot-shaped bulbs that give rise to an extensive fibrous root system. The large size and high wildlife value of this species have encouraged more nurseries to cultivate it, and gardeners are starting to catch on to the potential of bulb panicgrass as a landscaping plant that stands on its own as a focal specimen or blends well into mixed plantings of bird and mammal-friendly shrubs.

Vines

As a growth form, vines are more abundant in tropical environments, but the Southwest is home to a surprising diversity of species that will happily climb, clamber, and twine through tree limbs and along fence lines. Including these species in a landscape adds a whole other layer of texture and structure to native plant gardens, where they can obscure unsightly block walls and chain-link fences. This section also includes some spreading groundcovers that can be utilized to knit together barren ground both visually and literally, as their root systems bind soil particles together, helping to prevent erosion. Most of our vine and groundcover species do well in shade and can be a great asset in filling areas where shrubs or wildflowers struggle from lack of sunlight. The species in this section are excellent for wildlife as shelter, food, or host plants that will contribute to the natural aesthetic and wildlife value of your yard.

Anemopsis californica – Saururaceae

yerba mansa (hierba del mansa)

1′ × spreads

full sun, partial shade, shade

high water needs

flowers April–October

Field guide: Look for this species between 1000 and 6000 feet in wetlands and drainages in Arizona Upland Sonoran Desertscrub, Mojave Desertscrub, and Chihuahuan Desertscrub into Semidesert Grassland, Madrean Evergreen Woodland, and Great Basin Conifer Woodland. The permanently moist habitats where this species occurs are fast disappearing because of river channelization and greater demand for groundwater.

Description: This sole member of the genus *Anemopsis* is one of the Southwest's most storied medicinal plants, used by herbalists of every culture in our region to treat infections and inflammation. Its inflorescences are distinctive conical towers of tiny, cream-colored blossoms subtended by white bracts that become speckled with pink as they age. The flowers attract many pollinators. This species prefers wetlands, streams, bogs, and ciénagas and requires constant moisture to thrive; it's often planted in ponds and grows beautifully in hanging pots, gray-water outlets, and even aquaponics tanks. Pairs of fragrant oval leaves rise from a woody tuber and send out runners that form new leaves and roots where they touch the soil, helping this plant develop massive patches where conditions are favorable. Don't be afraid if the plant dies back from winter cold or summer heat—it can persist in the form of tubers and shoot up new leaves when the weather improves.

Aristolochia watsonii – Aristolochiaceae

southwestern pipevine (hierba del Indio)

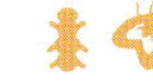

2–3″ × 6″–1′

full sun, partial shade, shade

low water needs

flowers April–October

Field guide: Can be found on loose, well-draining soil in Arizona Upland Sonoran Desertscrub and Chihuahuan Desertscrub and in Semidesert Grassland. This species occurs between 1000 and 5000 feet, usually in sandy or gravelly dry washes.

Description: So diminutive it can be easy to miss—and yet an essential element of a desert pollinator garden, where its toxic, arrowhead-shaped leaves will attract and feed pipevine swallowtail (*Battus philenor*) caterpillars. The musty-scented, urn-shaped flowers attract tiny flies that become trapped in the blooms by minute hairs that eventually wither, releasing the hapless insects to blunder into the next blossom in a somewhat devious pollination strategy. Plant pipevine in well-draining soil as a groundcover under other butterfly plants, like milkweeds (*Asclepias* spp.), mistflower (*Conoclinium dissectum*), and desert senna (*Senna covesii*), or put it into a hanging pot and let it drape over the edges. The wafer-like seeds readily germinate, so this species tends to spread, although its small size means you may not notice right away.

Cissus trifoliata – Vitaceae

grape ivy (hierba del buey)

30′ × 20′

partial shade, shade

medium water needs

flowers March–September

Field guide: In the Southwest, mostly restricted to Riparian Corridors where Arizona Upland Sonoran Desertscrub and Semidesert Grassland communities come together, around 2500 to 4500 feet. Typically found in well-draining sandy or gravelly soil where the tuber isn't likely to rot.

Description: This species occurs in scattered locations around the southern portions of Arizona and New Mexico and has probably benefited from human activity, as it tends to thrive in disturbed areas. Gardeners can appreciate this weedy quality because grape ivy is perfect for covering a fence line, enveloping block walls, or clambering up a tree. It grows from a potato-like tuber and sends up stems that adhere to seemingly any surface by means of threading tendrils. The fleshy, three-lobed leaves are caterpillar food for satellite (*Eumorpha satellitia*) and vine (*Eumorpha vitis*) sphinx moths, and the clusters of dark, grape-like berries are eaten by birds and mammals who will spread them around a landscape, especially under trees. This is one of our best vines for shaded areas, and it can quickly form dense thickets that may die back with winter frost but will return in spring with even more vigor.

Clematis ligusticifolia – Ranunculaceae

virgin's bower (barba de chivato)

20′+ × 20′+

full sun, partial shade, shade

medium water needs

flowers May–September

Field guide: Common in streamside thickets and canyons throughout the American West; in our region, it is encountered between 3000 and 8500 feet in Riparian Corridors running through Madrean Evergreen Woodland, Great Basin Conifer Woodland, and Montane Conifer Forest.

Description: A lovely and fast-growing vine with milky-white flowers emerging in clusters from the leaf axils. Although pollen- and seed-bearing blossoms occur on separate plants, both attract bees and butterflies. The pistillate flowers produce shaggy manes of long-tailed seeds that catch the wind and blow away to germinate in favorable locations. Extremely cold-tolerant and ideal for high-elevation gardens, but below 4500 feet in more arid habitats, it is replaced by Texas virgin's bower (*Clematis drummondii*), which is similar in appearance but better suited to the heat and harsh conditions of desert landscapes. Both species are great for growing up a trellis around a water-harvesting cistern, weaving into a fence line, or planting at the base of an arbor to create shade. Virgin's bower hosts caterpillars of the fatal metalmark butterfly (*Calephelis nemesis*) and Meske's pero moth (*Pero meskaria*).

Cottsia gracilis – Malpighiaceae

slender janusia (fermina)

6′ × 4′

full sun, partial shade

low water needs

flowers April–October

Field guide: Most often found on exposed rocky slopes in the Sonoran, Chihuahuan, and Mojave deserts into Semidesert Grassland communities, this plant may occur from 1000 up to 5000 feet at warmer sites.

Description: One of many engrossing Sonoran Desert botanical sights is that of a scraggly vine enveloping the spiny branches of an ocotillo (*Fouquieria splendens*), garlanded with small yellow flowers and buzzing with furry-legged bees. That's this vine: a tough, wiry plant with thin stems and pairs of narrow leaves that shift in color between dull green and dusky purple seasonally. When the petite yellow blossoms appear in spring, they exude an oily substance rich in proteins and fats that attracts female bees in the genus *Centris*, who use their squeegee-like hind legs to harvest the oil as a food source for their larvae. These bees are important pollinators of other species, including desert willow (*Chilopsis linearis*) and paloverdes (*Parkinsonia* spp.). Plant this vine for a front-row seat to the fascinating ecological show and to support populations of an important native pollinator.

Cucurbita foetidissima – Cucurbitaceae

buffalo gourd (calabaza de coyote)

1′ × 6–10′

full sun, partial shade

medium water needs

flowers May–August

Field guide: Most likely to be found in sandy fields or disturbed roadsides between 2500 and 7000 feet in Semidesert and Great Basin Grassland, this can also be seen in Chihuahuan Desertscrub, Interior Chaparral, and Great Basin Conifer Woodland among junipers (*Juniperus* spp.) and pines (*Pinus* spp.).

Description: Across a broad geographic and elevational range, buffalo gourd cascades down sandy wash edges and snakes along open grasslands. It grows from a woody tuber with thin stems that hug the ground and send up sail-shaped leaves interspersed with bell-like golden flowers, which are pollinated primarily by specialized squash bees. It is not uncommon to find male squash bees covered in pollen, seemingly drunk, and fast asleep inside the blossoms. The flowers give rise to spherical green-and-white striped fruits filled with bitter pulp. In the low desert, this species is often replaced with coyote gourd (*Cucurbita digitata*), which has similar blooms but hand-shaped leaves with narrow "fingers." Plant in well-draining soils since the tuber can be prone to rot in wintertime. Interplant with trees like mesquites (*Neltuma* spp.), paloverdes (*Parkinsonia* spp.), or oaks (*Quercus* spp.), and allow the vine to climb the branches. In fall, the fruits will turn yellow and hang like decorative ornaments.

Dichondra brachypoda – Convolvulaceae

New Mexico pony's foot (oreja de ratón)

2″ × 3′+

partial shade, shade

medium water needs

flowers July–October

Field guide: Found in shady groves and wetlands from Semidesert Grassland through Madrean Evergreen Woodland and Great Basin Conifer Woodland from 4500 and 6500 feet. It is not particularly common in habitat and exists in widely dispersed populations.

Description: One of our best groundcovers for shade, New Mexico pony's foot is gaining a lot of traction in the horticulture trade for its efficacy in rock gardens, hanging baskets, and shaded courtyards. The hoof-shaped leaves trail along, and the stem will root in at the nodes where it touches the soil, creating a dense but low cover. This species can be planted at the base of a Texas mulberry (*Morus microphylla*), kidneywood (*Eysenhardtia orthocarpa*), or western soapberry (*Sapindus saponaria* var. *drummondii*) in a large pot or raised bed, where the stems can cascade over the edges.

Humulus lupulus var. *neomexicanus* – Cannabaceae

hops (lúpulo)

20′+ × 20′+

partial shade, shade

high water needs

flowers July–August

Field guide: A species of wooded Riparian Corridor habitats in Montane Conifer Forest and Subalpine Conifer Forest, from as low as 6000 feet but generally between 7000 and 9500 feet. Getting to these plants often involves bushwhacking through dense riparian growth to find them clambering up cottonwoods (*Populus* spp.) or clinging to rocky slopes.

Description: Most of us will have tasted hops even if we have never seen it, due to its essential role in flavoring beer. But far from the hop yards of the Pacific Northwest and Bavaria, in the rugged canyons of the Southwest, our native hops forms luxurious patches clambering up cliff walls and winding into trees, producing those papery, cone-like flowers that have become such an important global commodity. The dense growth form and relatively large leaves of hops make it ideal for trellising against a wall or covering an arbor, though in lower-elevation gardens, it requires shade to thrive. A relatively high water requirement means that this plant will be happiest where consistent moisture is available from irrigation, gray water, and rain runoff, and hops will benefit from a mulch of wood chips or leaf material. Try using hops as a screen for a water-harvesting cistern or on the wall of an outdoor shower. The foliage is a food source for the hop vine moth (*Hypena humuli*) and red admiral butterfly (*Vanessa atalanta*).

Maurandya antirrhiniflora – Plantaginaceae

snapdragon vine (enredadera del cielo)

6–9′ × 6′

full sun, partial shade

low to medium water needs

flowers April–October

Field guide: Broadly distributed across the Southwest, from the sky islands of Arizona and New Mexico to the Rio Grande Valley in Texas and the northwestern part of Arizona (around the Grand Canyon) between 1500 and 6000 feet. Follow dry washes with loose rocks on the edges, and you may find this plant among dense shrubs or spilling down from a stone ledge.

Description: Rarely exceeding six feet in height, this vine likes to trail along boulders and slopes or twist among the branches of shrubs. In lower-elevation gardens, partial shade is recommended, but it will tolerate full sun in upland yards. This species is an ideal companion vine for velvet mesquite (*Neltuma velutina*) or oak (*Quercus* spp.), depending on where in the Southwest you are. The trumpet-shaped flowers can be purple, pink, or red, depending on the plant's provenance. Snapdragon vine produces abundant seed, and it can go feral in a landscape, popping up around shrubs and trees near the original planting. This species is a host plant for the gray and yellow caterpillars of the sympistis sokar moth (*Sympistis sorapis*).

Parthenocissus inserta – Vitaceae

thicket creeper (parra virgen)

25′ × 25′

full sun, partial shade

medium to high water needs

flowers May–June

Field guide: Usually seen in Riparian Corridors, where it favors moist canyon edges, disturbed fields, and wetlands in Great Basin Conifer Woodland and Montane Conifer Forest habitats. Look for it between 3000 and 7000 feet, often forming prodigious patches growing up fence posts, trees, and cliff faces. Typically found in association with cottonwoods (*Populus* spp.) and ashes (*Fraxinus* spp.).

Description: This vigorous vine forms thick sheets of foliage that sometimes cover huge expanses. A member of the grape family, it has glossy leaves that are hand-shaped (palmate) with serrations along the margins, and the foliage turns brilliant scarlet in fall. Several species of moths use the leaves as a caterpillar food source, including the picturesque achemon sphinx moth (*Eumorpha achemon*) and enigmatic underwing moths (*Catocala* spp.). Bees pollinate the nondescript green flowers, and the deep purple grape-like berries attract a host of bird species. Creeper vine requires ample moisture to take off, but if conditions are favorable, it can be a very aggressive grower, so be sure to give it space to spread.

Vitis arizonica – Vitaceae

canyon grape (uva del monte)

15′ × 10′

full sun, partial shade, shade

high water needs

flowers April–July

Field guide: Thrives in Riparian Corridors from the upper reaches of the Southwest's desert communities all the way up to Montane Conifer Forest habitats. May be found between 2000 and 7500 feet anywhere that willows (*Salix* spp.), ashes (*Fraxinus* spp.), and cottonwoods (*Populus* spp.) grow.

Description: This fast-growing wild grape rises from a woody base and forms a tangle of thin stems and deciduous, heart-shaped leaves. With the onset of summer, grapeleaf skeletonizers (*Harrisina americana*) and leafcutter bees (*Megachile* spp.) may munch on the foliage. You can remove them by hand or deter them using a soapy-water mixture, but if left alone, these caterpillars will become food for birds and disappear fairly quickly. The pollen- and fruit-bearing flowers are borne on separate individuals, so multiple plants are required for fruit to set. The grapes are quite small and range in flavor from exceedingly delicious to unbearably bitter, but all tend to have a tannic, chalky aftertaste. Plant canyon grape under a tree or at the base of a trellis around a water-harvesting tank; the growth rate is outstanding with reliable water.

Plant Guild Lists

THE SONORAN HOT SPOT

This is a guild for exposed south- or west-facing exposures in Sonoran Desert gardens in places like Phoenix, Yuma, and Tucson, Arizona. These plants can tolerate extreme heat and sun and, as they grow, will start to cool and shade a difficult garden space.

Trees

- *Olneya tesota* – ironwood
- *Parkinsonia microphylla* – foothill paloverde

Shrubs

- *Calliandra eriophylla* – fairy duster
- *Larrea tridentata* – creosote bush
- *Lycium berlandieri* – Berlandier's wolfberry
- *Simmondsia chinensis* – jojoba

Perennials

- *Baileya multiradiata* – desert marigold
- *Encelia farinosa* – brittlebush
- *Sphaeralcea ambigua* – apricot globemallow

Grasses

- *Muhlenbergia porteri* – bush muhly

Cacti and Succulents

- *Carnegiea gigantea* – saguaro
- *Ferocactus wislizeni* – fishhook barrel cactus
- *Fouquieria splendens* – ocotillo
- *Opuntia engelmannii* – Engelmann pricklypear

Vines

- *Cottsia gracilis* – slender janusia

THE HIGH CHAPARRAL

Residents of the Prescott and Payson areas in Arizona can assemble Interior Chaparral species to create a garden that imparts privacy and noise screening while attracting a diversity of wildlife.

Trees

- *Quercus arizonica* – Arizona white oak

Shrubs

- *Garrya wrightii* – Wright's silktassel
- *Purshia stansburyana* – cliffrose
- *Quercus turbinella* – scrub oak
- *Rhus ovata* – sugar sumac

Perennials

- *Melampodium leucanthum* – blackfoot daisy
- *Penstemon palmeri* – Palmer's penstemon

Grasses

- *Bouteloua curtipendula* – sideoats grama
- *Hesperostipa neomexicana* – New Mexico feathergrass

Cacti and Succulents

- *Nolina microcarpa* – beargrass
- *Opuntia engelmannii* – Engelmann pricklypear
- *Yucca baccata* – banana yucca

Vines

- *Clematis ligusticifolia* – virgin's bower

◀ Planting in guilds rather than situating single plants here and there helps mimic natural habitats and create diverse, layered plantings.

THE SKY ISLAND

Combining species from the Semidesert Grassland and Madrean Evergreen Woodland habitats found near the US–Mexico boundary, this guild creates a layered habitat that's perfect for backyard birdwatching.

Trees

- *Eysenhardtia orthocarpa* – kidneywood
- *Juniperus deppeana* – alligator juniper
- *Quercus emoryi* – Emory oak

Shrubs

- *Arctostaphylos pungens* – pointleaf manzanita
- *Capsicum annuum* var. *glabriusculum* – chiltepin
- *Mahonia haematocarpa* – red barberry
- *Mimosa dysocarpa* – velvetpod mimosa

Perennials

- *Glandularia gooddingii* – Goodding's mock vervain
- *Mandevilla brachysiphon* – Huachuca rock trumpet

Grasses

- *Bouteloua chondrosioides* – sprucetop grama
- *Elionurus barbiculmis* – woolyspike balsamscale
- *Eragrostis intermedia* – plains lovegrass

Cacti and Succulents

- *Agave palmeri* – Palmer's agave
- *Dasylirion wheeleri* – desert spoon
- *Yucca schottii* – mountain yucca

Vines

- *Cissus trifoliata* – grape ivy

CHIHUAHUAN BIRD AND BUG

This guild features a mix of desert and grassland species from the Chihuahuan Desert. The plants in this list will attract birds, including hummingbirds, while providing food for caterpillars and adults of various butterfly and moth species. This guild is well-suited to gardens in places like El Paso, Texas, and Las Cruces, New Mexico.

Trees

- *Neltuma glandulosa* – honey mesquite

Shrubs

- *Larrea tridentata* – creosote bush
- *Lycium pallidum* – pale wolfberry
- *Parthenium incanum* – mariola

Grasses

- *Aristida purpurea* – purple threeawn
- *Digitaria californica* – Arizona cottontop
- *Muhlenbergia porteri* – bush muhly
- *Sporobolus airoides* – alkali sacaton

Cacti and Succulents

- *Cylindropuntia imbricata* – tree cholla
- *Fouquieria splendens* – ocotillo
- *Yucca elata* – soaptree yucca

Vines

- *Maurandya antirrhiniflora* – snapdragon vine

GREAT BASIN GRASSLAND AND WOODLAND

With a combination of grassland and woodland species from the Great Basin, this guild creates a layered landscape that provides excellent nesting habitat and food for birds and mammals. These species will handle cold and heat with equal aplomb and are a good fit for cities like Albuquerque, New Mexico, and Holbrook, Arizona.

Trees

- *Juniperus monosperma* – oneseed juniper
- *Pinus edulis* – pinyon pine

Shrubs

- *Artemisia tridentata* – big sagebrush
- *Chamaebatiaria millefolium* – fernbush
- *Krascheninnikovia lanata* – winterfat
- *Rhus aromatica* var. *trilobata* – three-leaf sumac

Grasses

- *Achnatherum hymenoides* – Indian ricegrass
- *Bouteloua gracilis* – blue grama
- *Elymus trachycaulus* – slender wheatgrass

Cacti and Succulents

- *Echinocereus coccineus* – claret cup hedgehog
- *Opuntia polyacantha* – plains pricklypear
- *Yucca angustissima* – narrowleaf yucca

Vines

- *Cucurbita foetidissima* – buffalo gourd

COOL MOUNTAIN AIR

These plants are drawn from the conifer forests found on mountain ranges around the Southwest and can be used to create privacy and cover while attracting a range of pollinators. These species will thrive in cold, but their sensitivity to heat means this list is for high-elevation gardens only.

Trees

- *Acer negundo* – boxelder maple
- *Juniperus scopulorum* – Rocky Mountain juniper
- *Pinus ponderosa* – ponderosa pine

Shrubs

- *Cercocarpus montanus* – mountain mahogany
- *Robinia neomexicana* – New Mexico locust
- *Rosa woodsii* – Woods' rose
- *Ribes pinetorum* – orange gooseberry

Perennials

- *Achillea millefolium* – yarrow
- *Eriogonum umbellatum* – sulfur buckwheat
- *Mahonia repens* – creeping barberry

Cacti and Succulents

- *Agave parryi* – Parry's agave
- *Echinocereus coccineus* – claret cup hedgehog
- *Escobaria vivipara* – spinystar cactus

Vines

- *Vitis arizonica* – canyon grape

Bird Food and Shelter Plants

- *Abies concolor* – white fir
- *Abutilon* spp. – mallows
- *Acer* spp. – maples
- *Aloysia gratissima* – bee brush
- *Atriplex canescens* – fourwing saltbush
- *Bursera microphylla* – elephant tree
- *Capsicum annuum* var. *glabriusculum* – chiltepin
- *Carnegiea gigantea* – saguaro
- *Celtis* spp. – hackberries
- *Cercocarpus montanus* – mountain mahogany
- *Cissus trifoliata* – grape ivy
- *Condalia warnockii* – Warnock's snakewood
- *Cylindropuntia* spp. – chollas
- *Echinocereus* spp. – hedgehog cacti
- *Encelia farinosa* – brittlebush
- *Eriogonum* spp. – buckwheats
- *Fallugia paradoxa* – Apache plume
- *Forestiera pubescens* – stretchberry
- *Frangula californica* – California buckthorn
- *Fraxinus velutina* – velvet ash
- *Gaillardia pulchella* – firewheel
- *Garrya wrightii* – Wright's silktassel
- *Heliomeris multiflora* – showy goldeneye
- *Hesperocyparis arizonica* – Arizona cypress
- *Holodiscus discolor* – creambush
- *Juglans major* – black walnut
- *Juniperus* spp. – junipers
- *Krascheninnikovia lanata* – winterfat
- *Lycium* spp. – wolfberries
- *Lysiloma watsonii* – feather tree
- *Mahonia* spp. – barberries
- *Morus microphylla* – Texas mulberry
- *Neltuma* spp. – mesquites
- *Opuntia* spp. – pricklypears
- *Parkinsonia* spp. – paloverdes
- *Parthenocissus inserta* – thicket creeper
- *Picea pungens* – blue spruce
- *Pinus* spp. – pines
- *Platanus wrightii* – Arizona sycamore
- *Populus* spp. – cottonwood and aspen
- *Poaceae* spp. – grasses
- *Quercus* spp. – oaks
- *Rhus* spp. – sumacs
- *Ribes* spp. – currants
- *Rosa woodsii* – Woods' rose
- *Sambucus* spp. – elderberries
- *Sapindus saponaria* var. *drummondii* – western soapberry
- *Sarcomphalus obtusifolius* – graythorn
- *Sphaeralcea* spp. – globemallows
- *Thymophylla pentachaeta* – golden fleece
- *Vitis arizonica* – canyon grape

Butterfly Nectar and Larval Plants

- Almost every plant in this book!

Hummingbird Plants

- *Abutilon* spp. – mallows
- *Agave* spp. – agaves
- *Aquilegia desertorum* – desert columbine
- *Anisacanthus thurberi* – desert honeysuckle
- *Bouvardia ternifolia* – firecracker bush
- *Campanula rotundifolia* – bluebell bellflower
- *Chilopsis linearis* – desert willow
- *Echinocereus coccineus* – claret cup hedgehog
- *Epilobium canum* – hummingbird trumpet
- *Eriogonum fasciculatum* – flattop buckwheat
- *Erythranthe cardinalis* – scarlet monkeyflower
- *Fouquieria splendens* – ocotillo
- *Hedeoma nana* – dwarf false pennyroyal
- *Holodiscus discolor* – creambush
- *Ipomopsis aggregata* – scarlet gilia
- *Iris missouriensis* – Rocky Mountain iris
- *Justicia californica* – chuparosa
- *Lycium* spp. – wolfberries
- *Mandevilla brachysiphon* – Huachuca rock trumpet
- *Maurandya antirrhiniflora* – snapdragon vine
- *Menodora scabra* – rough menodora
- *Mirabilis* spp. – four o'clocks
- *Parkinsonia* spp. – paloverdes
- *Penstemon* spp. – beardtongues
- *Plumbago zeylanica* – leadwort
- *Poliomintha incana* – frosted mint
- *Ruellia ciliatiflora* – violet wild petunia
- *Salvia arizonica* – Arizona sage
- *Silene laciniata* – cardinal catchfly
- *Tecoma stans* – yellow bells

Desert Tortoise Food Plants

- *Abutilon* spp. – mallows
- *Acaciella angustissima* – prairie acacia
- *Anisacanthus thurberi* – desert honeysuckle
- *Calliandra eriophylla* – fairy duster
- *Calylophus hartwegii* – Hartweg's sundrops
- *Carnegiea gigantea* – saguaro
- *Chilopsis linearis* – desert willow
- *Cissus trifoliata* – grape ivy
- *Cochemiea/Mammillaria* spp. – pincushion cacti
- *Commelina erecta* – whitemouth dayflower
- *Dalea* spp. – prairie clovers
- *Dicliptera resupinata* – Arizona foldwing
- *Echinocereus* spp. – hedgehog cacti
- *Eriogonum* spp. – buckwheats
- *Fouquieria splendens* – ocotillo
- *Hibiscus* spp. – hibiscus
- *Justicia californica* – chuparosa
- *Larrea tridentata* – creosote bush
- *Maurandya antirrhiniflora* – snapdragon vine
- *Mirabilis* spp. – four o'clocks
- *Oenothera* spp. – primrose
- *Opuntia* spp. – pricklypears
- *Poaceae* spp. – grasses
- *Ruellia ciliatiflora* – violet wild petunia
- *Senna covesii* – desert senna
- *Sphaeralcea* spp. – globemallows
- *Tecoma stans* – yellow bells
- *Thymophylla pentachaeta* – golden fleece

Resources

Online Resources

The Arizona Native Plant Society
aznps.com

Audubon Southwest
southwest.audubon.org/about-us/where-we-are

Bug Guide
bugguide.net/node/view/15740

Butterflies and Moths of North America
butterfliesandmoths.org

California Native Plant Society
cnps.org

Colorado Native Plant Society
conps.org

Cornell Lab of Ornithology
birds.cornell.edu/home

Desert Survivors Inc. (Plant Lists)
desertsurvivors.org/plants

Flora of North America
floranorthamerica.org/Main_Page

Lady Bird Johnson Wildflower Center
wildflower.org

Nevada Native Plant Society
nvnps.org

Native Plant Society of New Mexico
npsnm.org

Native Plant Society of Texas
npsot.org

Reducing the Urban Heat Island Effect with Native Trees in Tucson, Arizona
repository.arizona.edu/bitstream/handle/10150/664136/SBE_2022_Eason_Capstone_Poster.pdf?sequence=5&isAllowed=y

Southwest Biodiversity (SEINet)
swbiodiversity.org/seinet/index.php

Southwest Desert Flora
southwestdesertflora.com

Spadefoot Nursery (Plant Info)
spadefootnursery.com/plant-information

USDA PLANTS Database
plants.usda.gov/home

Utah Native Plant Society
unps.org

Wildflowers, Ferns, & Trees of Colorado, New Mexico, Arizona, & Utah
swcoloradowildflowers.com/index.htm

Books

Natural History

Breslin, Peter, Rob Romero, Greg Starr, and Vonn Watkins. 2017. *Field Guide to Cacti & Other Succulents of Arizona*. 2nd ed. Tucson, AZ: Tucson Cactus and Succulent Society.

Brock, Jim P. 2008. *Butterflies of the Southwest*. Tucson, AZ: Rio Nuevo Publishers.

Brown, David E. 1994. *Biotic Communities: Southwestern United States and Northwestern Mexico*. Salt Lake City, UT: University of Utah Press.

Burns, Jim. 2008. *Jim Burns' Arizona Birds: From the Backyard to the Backwoods*. 3rd ed. Tucson, AZ: The University of Arizona Press.

Carter, Jack L. 2012. *Trees and Shrubs of New Mexico*. 2nd ed. Silver City, NM: Mimbres Press.

Carter, Jack L., Martha A. Carter, and Donna J. Stevens. 2003. *Common Southwestern Native Plants: An Identification Guide*. Silver City, NM: Mimbres Press.

Chronic, Halka. 1983. *Roadside Geology of Arizona*. Missoula, MT: Mountain Press Publishing Company.

Epple, Anne Orth, Lewis E. Epple, and John F. Wiens. 2021. *Plants of Arizona*. 3rd ed. Guilford, CT: Falcon Guides.

Felger, Richard Stephen, James Thomas Verrier, Kelly Kindscher, and Xavier Raj Herbst Khera. 2021. *Field Guide to the Trees of the Gila Region of New Mexico*. Albuquerque, NM: University of New Mexico Press.

Gould, Frank W. 1951. *Grasses of Southwestern United States*. Tucson, AZ: The University of Arizona Press.

Johnson, Matthew Brian. 2004. *Cacti, Other Succulents, and Unusual Xerophytes of Southern Arizona*. Superior, AZ: Boyce Thompson Arboretum.

Kaufman, Lynn Hassler. 2002. *Hummingbirds of the American West*. Tucson, AZ: Rio Nuevo Publishers.

Kearney, Thomas H., and Robert H. Peebles. 1960. *Arizona Flora*. 2nd ed. Berkeley, CA: University of California Press.

Koweek, Jim. 2016. *Grassland Plant ID for Everyone*. Patagonia, AZ: Rafter Lazy K Publishing.

Lanner, Ronald M. 1984. *Trees of the Great Basin: A Natural History*. Reno, NV: University of Nevada Press.

Mozingo, Hugh N. 1987. *Shrubs of the Great Basin: A Natural History*. Reno, NV: University of Nevada Press.

Turner, Raymond M., Janice E. Bowers, and Tony L. Burgess. 1995. *Sonoran Desert Plants: An Ecological Atlas*. Tucson, AZ: The University of Arizona Press.

Werner, Floyd and Carl Olson. 1994. *Learning about and Living with Insects of the Southwest*. Tucson, AZ: Fisher Books.

Gardening

Busco, Janice and Nancy R. Morin. 2003. *Native Plants for High-Elevation Western Gardens*. Golden, CO: Fulcrum Publishing.

Chance, Leo J. 2012. *Cacti and Succulents for Cold Climates: 274 Outstanding Species for Challenging Conditions*. Portland, OR: Timber Press.

Johnson, Noelle. 2023. *Dry Climate Gardening: Growing Beautiful, Sustainable Gardens in Low-Water Conditions*. Beverly, MA: Cool Springs Press.

Lancaster, Brad. 2013. *Rainwater Harvesting for Drylands and Beyond*. 2nd ed. Tucson, AZ: Rainsource Press.

Lowenfels, Jeff and Wayne Lewis. 2010. *Teaming With Microbes: The Organic Gardener's Guide to the Soil Food Web*. Revised ed. Portland, OR: Timber Press.

Mielke, Judy. 1993. *Native Plants for Southwestern Landscapes*. Austin, TX: University of Texas Press.

Miller, George Oxford. 2007. *Landscaping with Native Plants of the Southwest*. Minneapolis, MN: Voyageur Press.

Miller, George Oxford. 2021. *Native Plant Gardening for Birds, Bees & Butterflies: Southwest*. Cambridge, MN: Adventure Publications.

Nokes, Jill. 1986. *How to Grow Native Plants of Texas and the Southwest*. Austin, TX: Texas Monthly Press.

Starr, Greg. 2012. *Agaves: Living Sculptures for Landscapes and Containers*. Portland, OR: Timber Press.

Starr, Greg. 2009. *Cool Plants for Hot Gardens: 200 Water-Smart Choices for the Southwest*. Tucson, AZ: Rio Nuevo Publishers.

Tallamy, Douglas W. 2020. *Nature's Best Hope: A New Approach to Conservation that Starts in Your Yard*. Portland, OR: Timber Press.

Acknowledgments

This book has been made possible by the help and encouragement of a small army of people. Grandma Dot and Grandpa Richard sparked a love of the outdoors in an obstreperous child. The Swenson and Lukins families constantly supported our various (mis)adventures. Anna Brody, Michael Sadatmousavi, and Sue Carnahan contributed their lovely photographs. The board and members of The Arizona Native Plant Society helped shepherd this work, and special thanks go to Doug Ripley and all the members who assisted with reviewing and editing. Appreciation also goes to the Native Plant Society of New Mexico and their fearless leaders Don and Wendy Graves. We have been privileged to visit numerous public and private gardens throughout the Southwest and want to extend our heartfelt gratitude to each and every gardener who shared their experience and time with us. In particular, we are grateful to the horticultural staff at Desert Survivors native plant nursery and the Arizona–Sonora Desert Museum for imparting so much wisdom. A huge thank you to Ryan Harrington, Sarah Milhollin, Ellen Foreman, and the team at Timber Press for their guidance and work on making this project a reality. A special nod is due to Janice Busco for being there at the genesis of this book and influencing the plant selection. Finally, thank you to Lano, Julian, and Reign for making it all possible.

Photography Credits

All photographs are by Luke Takata unless otherwise stated.

Anna Brody, page 252.

Sue Carnahan, page 192.

Michael Sadatmousavi, pages 38 (bottom left), 39 (right), 40 (bottom), 70 (top left, bottom right).

Mary Velgos, page 170 (bottom left)

Index

© Anna Brody

© Luke Takata

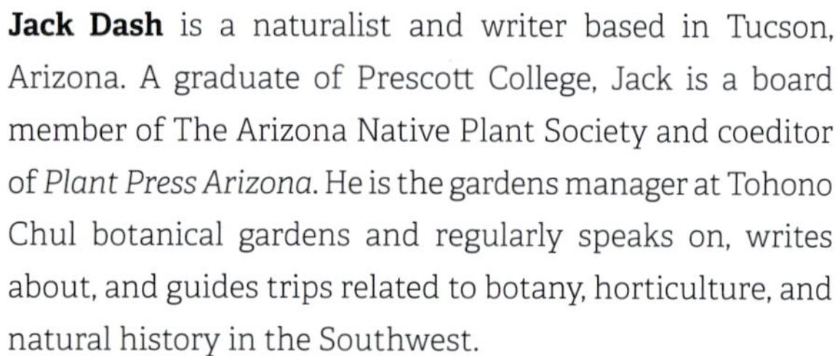

Jack Dash is a naturalist and writer based in Tucson, Arizona. A graduate of Prescott College, Jack is a board member of The Arizona Native Plant Society and coeditor of *Plant Press Arizona*. He is the gardens manager at Tohono Chul botanical gardens and regularly speaks on, writes about, and guides trips related to botany, horticulture, and natural history in the Southwest.

Luke Takata is a photographer and storyteller based in Tucson, Arizona. A graduate of Pratt Institute and the International Center of Photography, his work has been recognized with an Award of Excellence from the Alexia Foundation, and his photographs are a part of the permanent collection at The Wittliff. Luke is the engagement manager at Tohono Chul botanical gardens.

Luke and Jack are the co-creators of Atascosa Borderlands, a visual storytelling project dedicated to a remote 42-mile stretch of the US–Mexico border in southern Arizona. You can find out more about their work by visiting: atascosaborderlands.com.